# WRITING ON THE WALL

BY THE SAME AUTHOR

*An Edible History of Humanity*

*A History of the World in 6 Glasses*

*The Turk*

*The Neptune File*

*The Victorian Internet*

# WRITING ON THE WALL

*Social Media – The First 2,000 Years*

TOM STANDAGE

BLOOMSBURY
LONDON • NEW DELHI • NEW YORK • SYDNEY

First published in Great Britain 2013



Bloomsbury Publishing plc
50 Bedford Square
London WC1B 3DP

www.bloomsbury.com

Bloomsbury Publishing, London, New Delhi, New York and Sydney

A CIP catalogue record for this book is available from the British Library

ISBN 978 1 4088 4206 5

10 9 8 7 6 5 4 3 2 1

Typeset by Westchester Book Group
Printed and bound in Great Britain by CPI Group (UK) Ltd, Croydon CR0 4YY

For Miles

CONTENTS


| | | |
|---|---|---|
| | Introduction: Cicero's Web | 1 |
| 1 | The Ancient Foundations of Social Media: Why Humans Are Wired for Sharing | 6 |
| 2 | The Roman Media: The First Social-Media Ecosystem | 21 |
| 3 | How Luther Went Viral: The Role of Social Media in Revolutions (1) | 48 |
| 4 | Poetry in Motion: Social Media for Self-Expression and Self-Promotion | 64 |
| 5 | Let Truth and Falsehood Grapple: The Challenges of Regulating Social Media | 82 |
| 6 | And So to the Coffeehouse: How Social Media Promotes Innovation | 104 |
| 7 | The Liberty of Printing: The Role of Social Media in Revolutions (2) | 124 |
| 8 | The Sentinel of the People: Tyranny, Optimism, and Social Media | 147 |
| 9 | The Rise of Mass Media: The Centralization Begins | 170 |
| 10 | The Opposite of Social Media: Media in the Broadcast Era | 189 |

11 The Rebirth of Social Media: From ARPANET to Facebook 214

Epilogue: History Retweets Itself 240

Acknowledgments 251

Notes 253

Sources 259

Index 269

# INTRODUCTION: CICERO'S WEB

*Not to know what has been transacted in former times is to be always a child. If no use is made of the labors of past ages, the world must remain always in the infancy of knowledge.*
—Marcus Tullius Cicero

IN JULY 51 B.C. the Roman statesman and orator Marcus Tullius Cicero arrived in Cilicia, in what is now southeast Turkey, to take up the post of proconsul, or regional governor. Cicero had been deeply reluctant to leave the bustle of Rome, where he was a central figure in the plotting and counterplotting of Roman politics, and he intended to return as soon as was decently possible. The burning question of the day was whether Julius Caesar, commander of Rome's armies in the west, would make a grab for power by marching on the city. Cicero had spent his career trying to defend the political system of the Roman republic, with its careful division of powers and strict limits on the authority of any individual, from Caesar and others who wished to centralize power and seize it for themselves. But a new anticorruption law required Cicero and other trustworthy elder statesmen to take up posts as provincial governors. Fortunately, even in distant Cilicia, Cicero had the means to stay in touch with the goings-on in Rome—because the Roman elite had developed an elaborate system to distribute information.

At the time there were no printing presses and no paper. Instead, information circulated through the exchange of letters and other documents which were copied, commented on, and shared with others in the form of papyrus rolls. Cicero's own correspondence, one of the best-preserved collections of letters from the period, shows that he exchanged letters constantly with his friends elsewhere, keeping them up to date with the latest political machinations, passing on items of interest from others, and providing his own commentary and opinions. Letters were often copied, shared, and quoted in other letters. Some letters were addressed to several people and were written to be read aloud, or to be posted in public for general consumption.

When Cicero or another politician made a noteworthy speech, he could distribute it by making copies available to his close associates, who would read it and pass it on to others. Many more people might then read the speech than had heard it being delivered. Books circulated in a similar way, as sets of papyrus rolls passed from person to person. Anyone who wished to retain a copy of a speech or book would have it transcribed by scribes before passing it on. Copies also circulated of the *acta diurna* (the "daily acts," or state gazette), the original of which was posted on a board in the Forum in Rome each day and contained summaries of political debates, proposals for new laws, announcements of births and deaths, the dates of public holidays, and other official information. As he departed for Cilicia, Cicero asked his friend and protégé Marcus Caelius Rufus to send him copies of each day's gazette along with his letters. But this would be just part of Cicero's information supply. "Others will write, many will bring me news, much too will reach me even in the way of rumor," Cicero wrote.

With information flitting from one correspondent to another, this informal system enabled information to penetrate to the farthest provinces within a few weeks at most. News from Rome took around five weeks to reach Britain in the west and seven weeks to reach Syria in the east. Merchants, soldiers, and officials in distant parts would circulate information from the heart of the republic within their own social circles, sharing extracts from letters, speeches, or the state gazette with their friends and passing news and rumors from the frontier back to their contacts in Rome. There was no formal postal service, so letters had to be carried by messengers or given to friends, traders, or travelers heading in the right direction. The result was that Cicero, along with other members of the Roman elite, was kept informed by a web of contacts—the members of his social circle—all of whom gathered, filtered, and distributed information for each other.

To modern eyes this all seems strangely familiar. Cicero was, to use today's Internet jargon, participating in a "social media" system: that is, an environment in which information was passed from one person to another along social connections, to create a distributed discussion or community. The Romans did it with papyrus rolls and messengers; today hundreds of millions of people do the same things rather more quickly and easily using Facebook, Twitter, blogs, and other Internet tools. The technologies involved are very different, but these two forms of social media, separated by two millennia, share many of the same underlying structures and dynamics: they are two-way, conversational environments in which information passes horizontally from one person to another along social networks, rather than being delivered vertically from an impersonal central source.

Cicero's web is just one of many historical antecedents of today's social media. Other notable examples include the circulation of letters and other documents in the early Christian church; the torrent of printed tracts that circulated in sixteenth-century Germany at the start of the Reformation; the exchange and copying of gossip-laden poetry in the Tudor and Stuart courts; the dueling political pamphlets with which Royalists and Parliamentarians courted public opinion during the English Civil War; the stream of news sheets and pamphlets that coursed through Enlightenment coffeehouses; the first scientific journals and correspondence societies, which enabled far-flung scientists to discuss and build upon each other's work; the pamphlets and local papers that rallied support for American independence; and the handwritten poems and newsletters of prerevolutionary France, which spread gossip from Paris throughout the country. Such social-media systems arose frequently because, for most of human history, social networks were the dominant means by which new ideas and information spread, in either spoken or written form. Over the centuries, the power, reach, and inclusivity of these social-media systems steadily increased.

But then, starting in the mid-nineteenth century, everything changed. The advent of the steam-powered printing press, followed in the twentieth century by radio and television, made possible what we now call "mass media." These new technologies of mass dissemination could supply information directly to large numbers of people with unprecedented speed and efficiency, but their high cost meant that control of the flow of information became concentrated in the hands of a select few. The delivery of information assumed a one-way, centralized, broadcast pattern that overshadowed the tradition of two-way, conversational, and social distribution that had come before. Vast media empires grew up around these mass-media technologies, which also fostered a sense of national identity and allowed autocratic governments to spread propaganda more easily than ever before.

In the past decade, however, the social nature of media has dramatically reasserted itself. The Internet has enabled a flowering of easy-to-use publishing tools and given social media unprecedented reach and scale, enabling it to compete with broadcast media and emerge from its shadow. The democratization of publishing made possible by Facebook, Twitter, YouTube, and other social platforms has been hugely disruptive for mass-media companies and, more importantly, is beginning to have far-reaching social and political effects. The re-emergence of social media, now supercharged by digital networks, represents a profound shift not just within the media, but within society as a whole.

And it has raised a host of difficult questions. Have new forms of social media led to a trivialization and coarsening of public discourse? How should those in authority respond when they face criticism in social media? Does social media inherently promote freedom and democracy? What is the role of social media, if any, in triggering revolutions? Is it a distracting waste of time that prevents people doing useful work? Is the use of social media actually antisocial, as online connections displace real-world interaction? Is social media just a fad that can be ignored?

This book will look for answers to these questions by considering a series of social-media systems that arose in very different times and places, but are linked by the common thread that they are based on the person-to-person sharing of information. These early forms of social media were involved in many of history's great revolutions. Concerns about the trivialization of public discourse, and the belief that new forms of media are dangerously distracting, go back centuries, as do the debates about the regulation of social-media systems and their ability to bring about social and political change. By examining the analog antecedents of today's digital social media, we can use history to cast new light on modern debates. At the same time, our modern experience of social media enables us to see the past with new eyes. Historical figures including Saint Paul, Martin Luther, and Thomas Paine are revealed as particularly adept users of social-media systems, with consequences that reverberate to this day.

All this will come as a surprise to modern Internet users who may assume that today's social-media environment is unprecedented. But many of the ways in which we share, consume, and manipulate information, even in the Internet era, build upon habits and conventions that date back centuries. Today's social-media users are the unwitting heirs of a rich tradition with surprisingly deep historical roots. Exposing these ancient precursors and tracing the story of the rise, fall, and rebirth of social media over the past two thousand years provides an illuminating new perspective on the history of Western media. It reveals that social media does not merely connect us to each other today—it also links us to the past.

# CHAPTER 1

# THE ANCIENT FOUNDATIONS OF SOCIAL MEDIA: WHY HUMANS ARE WIRED FOR SHARING

*Without gossip, there would be no society.*
—Robin Dunbar

## IT'S A SOCIAL WORLD

HOW LONG AGO did you last check Facebook? There's a good chance it was earlier today. The world's most popular social-networking site has more than one billion users, half of whom access it daily and a quarter of whom check it five or more times a day. Accessing social-networking sites is now the single most popular online activity worldwide: four out of five Internet users, or around 1.4 billion people, use social sites of one kind or another to post status updates, share photos and links, leave comments, and engage in discussions. Collectively, such sites account for a quarter of all time spent online globally, and more than 40 percent in some countries.

Facebook is the current leader of a huge international pack that includes Twitter, Google+, Tumblr, and LinkedIn, to name only a few companies that, like Facebook, are based in the United States. There are also strong players in other countries whose names may be less familiar: Qzone, Tencent Weibo, and Sina Weibo in China, Cyworld and me2day in South Korea, Orkut in Brazil, VKontakte and Odnoklassniki in Russia, Tuenti in Spain, and so on. Such is Facebook's dominance, however, that it alone accounts for one in seven minutes spent online around the world. Each month people collectively spend around three hundred billion minutes, or the equivalent of six hundred thousand years, on Facebook. That is not bad for a website that was only founded in 2004, and only opened its doors to nonacademic users in 2006.

The decline of MySpace, the previous industry leader, is a reminder that Facebook's continued dominance is by no means assured. But whoever is on top, it is clear that social networking sites have become a routine part of daily life for hundreds of millions of people, and an almost universal aspect of Internet use. Young people were the earliest adopters, but since 2010 the over-fifty-fives have caught up. In Britain and America, social sites of some sort are used by 98 percent of all Internet users, and the figure is above 90 percent in many other countries. The young tend to use social sites mostly to communicate with their friends; the old to stay in touch with their families.

The various social sites work in slightly different ways. Some require social connections between users to be approved by both parties, while others do not. Some assume that items posted are public, while others allow items to be shared only with specific individuals or groups. Some sites are intended for the sharing of particular types of content: Flickr for photos, SoundCloud for sound clips, YouTube for video. What they all have in common, however, is that they allow information to be shared along social networks with friends or followers (who may then share items in turn), and they enable discussion to take place around such shared information. Users of such sites do more than just passively consume information, in other words: they can also create it, comment on it, share it, discuss it, and even modify it. The result is a shared social environment and a sense of membership in a distributed community. What makes doing all this so enjoyable and compelling, and therefore so popular?

The answer has several components, all of which have deep behavioral and historical roots. The first, and most fundamental, is that as primates, humans are inherently social animals. Primate brains appear to have evolved specifically to process social information, to enable primates to function more effectively in groups. Second, one of the main ways humans assess and maintain their positions within social networks is by exchanging information with and about others (i.e., gossip). Through the exchange of gossip, individuals can advertise their status within the group and demonstrate their expertise, trustworthiness, and suitability as an ally or mate. Humans are, in short, built to form networks with others and to exchange information with them. The third component, media technology, starting with the emergence of writing, enables literate humans to extend this exchange of information across time and space to include people who are not physically present. The Internet, with its instant, global reach, does this particularly effectively, allowing users to share information with unprecedented ease. But it is by no means the first technology to have supported such a social-media environment; it is merely the most recent and most efficient way that humans have found to scratch a prehistoric itch.

The compelling nature of social media, then, can be traced back in part to the evolution of the social brain, as monkeys and other primates evolved over the past thirty-five million years; in part to the exchange of gossip following the emergence of human language, around one hundred thousand years ago; and in part to the origins of writing, around five thousand years ago. These are the three ancient foundations on which the social sharing of media, whether using papyrus scrolls in Roman times or the Internet today, has rested over the past two millennia. Let us consider each in turn.

## THE EVOLUTION OF THE SOCIAL BRAIN

There is something unusual about primates in general, and humans in particular. Compared with other animals, they have strikingly large brains relative to their bodies. Moreover, most of the extra brain volume is devoted to one part of the brain in particular: the neocortex, which is involved in higher functions such as spatial reasoning, sensory perception, and conscious thought. In most mammals the neocortex accounts for 30 to 40 percent of brain volume, but the proportion is over 65 percent in many primates, and 80 percent in humans. What is this large neocortex for?

One possibility is that primates evolved larger brains to enable them to use more complex tools, or to improve their ability to solve problems when searching for food, by doing things like cracking open nuts or extracting termites from their nests. The problem with this theory is that some primates with relative small neocortices (such as the aye-aye, a type of lemur found in Madagascar) display such "extractive foraging" behavior, while others with much larger neocortices (such as macaques) do not. And many nonprimates with much smaller neocortices, such as crows, are capable of solving complex problems. So that cannot be what the neocortex is for.

Another theory is that the large neocortex might have evolved to allow primates to build larger mental maps of their surroundings, to improve their ability to find food. But there turns out to be no correlation between neocortex volume and the range over which primates forage, or the average distance they travel each day. Nor do fruit-eating primates, which must constantly keep track of dispersed, short-lived supplies of food, have larger neocortices than leaf-eaters. So the neocortex was evidently not evolved for mapping the physical environment, either.

The odd thing is that all primates, not just those that use tools or solve complex problems, have disproportionately large brains. A large brain is expensive to develop and maintain. An adult human brain accounts for 2 percent of body mass, on average, but consumes around 20 percent of total energy intake. So there must something valuable that primates do that requires lots of extra mental processing capacity.

Along with their large brains, another distinguishing feature of primates is their social nature: they live in groups and have unusually complex social systems. They can form coalitions with their peers, for example, and are capable of deliberate deception, which requires the ability to hypothesize about another individual's view of the world. Living in a group is safer than living alone, because there are more eyes to spot predators and more hands to fend off rivals. But individual members must be able to balance their own needs with those of the group as a whole, rather than just looking out for themselves. Group members have to cooperate with others, understanding and anticipating their needs, while also establishing and managing their own positions within the ever-shifting pattern of alliances within the group.

In primates, these alliances are maintained through a process known as social grooming, which is carried out in pairs or small subgroups called coalitions. At its simplest level, this involves removing insects, parasites, or dirt from another individual's fur. But primates spend far more time grooming than is necessary for purely hygienic reasons: for some species, as much as 20 percent of their waking hours.

They do so in part because grooming is a pleasurable activity. Being groomed causes the release of beta-endorphins, which are natural opiates produced by the brain, resulting in a lower heart rate, a reduction in nervous behavior such as scratching, and a pleasant sense of relaxation. After being groomed, an individual will usually return the favor. Primates use social grooming to build strong bonds with a few other members of their group. They can also send social signals through the choices they make about which individuals they groom, how long they do it for, and which other individuals they allow to watch.

The time spent grooming is a worthwhile investment, because members of a grooming coalition will then support each other in several ways. They may gently steer a member of their coalition away from rivals in the wider group in order to minimize stress. If an individual is threatened by another group member, members of his grooming coalition will come to his aid. An individual can then face down an adversary, provided he has powerful enough allies to call upon. Group members may switch their allegiance from one coalition to another within the group if they believe it will improve their access to food, mating partners, or other resources. The constant interplay between grooming coalitions helps resolve and prevent conflicts, knitting the group as a whole together and making all its members safer from predators.

But tracking the relationships and alliances within the group, and evaluating the risks and rewards of aiding others when conflict arises, requires a lot of brainpower. In particular, it requires primates to theorize about how other members of their group feel toward each other and what their desires or intentions might be as a result. The larger the group, the more mental processing capacity is needed to keep track of the growing web of relationships. According to the "social-brain" theory, it was this need to analyze relationships within social networks, in order to support larger and therefore safer groups, that drove the evolution of larger and larger primate brains.

The theory is supported by the striking correlation, across a range of primate species, between neocortex size (as a percentage of overall brain volume) and group size, something that was first pointed out in 1992 by Robin Dunbar, a British anthropologist now at the University of Oxford. In howler monkeys, for example, the average group size is eight and the neocortex accounts for 65 percent of total brain

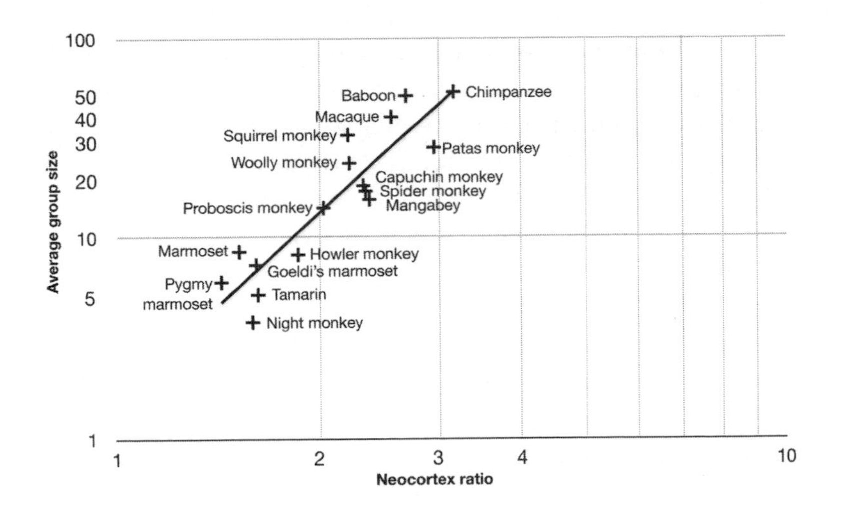


This chart shows the correlation between neocortex ratio (its size relative to the rest of the brain) and average group size in primates, first pointed out by Robin Dunbar. *Based on data from Dunbar, "Neocortex Size as a Constraint on Group Size in Primates" (1992)*

volume. For proboscis monkeys, the group size is fourteen and the neocortex volume is 67 percent; for capuchin monkeys, the figures are eighteen and 70 percent; for macaques, forty and 72 percent; for baboons, fifty-one and 73 percent; for chimpanzees, fifty-four and 76 percent. The fact that group size is strongly correlated with neocortex volume suggests that the primate brain is indeed a primarily social organ.

Further evidence for the social-brain theory comes from studies that compare neocortex size with deception rates in primates. Monkeys who discover a tasty food source, for example, may keep other members of their group away by feigning a lack of interest. And a young baboon about to be reprimanded by his mother may leap up and scan the horizon, tricking the rest of his troop into worrying that a rival troop is approaching and using this distraction to avoid punishment. Frequency of deception also turns out to be closely correlated with neocortex volume, supporting the idea that the benefit of a larger neocortex in primates is that it allows more elaborate social analysis and manipulation. Human brains are social brains, tuned to analyze the shifting intentions and allegiances of friends and rivals within a group. Our brains were literally made for social networking.

### TO SHARE IS HUMAN

How does all this apply to modern humans? Unlike other primates, we no longer live in small, roaming groups, and we do not spend hours each day picking parasites out of our friends' hair. Yet the modern equivalents of the social groups in which primate brains evolved, and the grooming behavior that bound them together, can be found right under our noses. When Dunbar analyzed the brain sizes and group sizes for apes he concluded that, given the size of the human neocortex, the average group size for humans should be 148, which he rounded to 150. This number, which has become known as the "Dunbar number," does indeed seem to recur frequently in human societies. It is the average population of a hunter-gatherer clan, of the earliest farming settlements in the ancient Near East, and of many villages recorded in the Domesday Book, a survey carried out in England in 1086.

More fundamentally it is, Dunbar believes, the largest group size in which it is possible for everyone to know everyone else. Above that size, some people will be strangers to others. It is therefore the maximum number of people with whom it is possible to have a reciprocal personal relationship: you know them well enough that they would come to your aid if needed, and you would do the same for them. This may explain why the Hutterites, a community of Christians who live in rural communes, have long chosen to split their communities when they exceed 150 people. They argue that maintaining order in a group any larger than that requires a police force; but below the 150-person limit, order can be maintained by peer pressure alone, because everyone knows each other. The Dunbar number is also the typical size of a military company, which generally includes between 120 and 180 individuals. A company in which everyone knows everyone else is a much more effective fighting unit.

The vast majority of Facebook users also turn out to have between 120 and 130 friends. Of course, some Facebook users have collected many more online "friends" than that. But they are more likely to be casual contacts than genuine friends. Most people, Dunbar's research has found, have five intimate real-world friends (akin to the members of a grooming coalition), and another ten close friends, within their larger network of 150. Interaction on Facebook (in the form of regular comments and messages) is similarly concentrated within a core group of intimates, with an average of seven other people for male users and ten people for female users. This core group is the digital equivalent of a grooming coalition.

But for humans, grooming is no longer a primarily physical activity. Instead, at some point in prehistory humans shifted away from physical forms of grooming and began to groom each other in another way: through speech, and specifically the exchange of "social information," or gossip, about other members of their social group. As with physical grooming, taking the time to have a chat with someone is a way to establish or strengthen a social bond. It also demonstrates the existence of that bond to others. But speech has three big advantages over physical forms of grooming. It allows grooming of more than one person at a time, while chatting in a small group. Grooming can also be carried out while performing another activity, such as eating, foraging for food, or resting. And grooming via speech, in the form of the exchange of gossip, enables people to find out about events within their social circle that they did not witness directly. This provides more information on which to base judgments about whether someone is trustworthy or not. And by passing on information selectively it is possible to manipulate one person's opinion of another. People can also form judgments about someone's trustworthiness by evaluating the accuracy of the information he or she passes on about others. Gossip is an extraordinarily rich source of social intelligence, both about the person speaking and about whoever is being discussed. And because our brains are wired to process just this kind of information, we find exchanging it extraordinarily compelling.

Such chatter benefits both the members of a group and the group as a whole. Individuals can better keep track of shifting alliances within the group, and passing accurate or useful information to others can help establish one's credibility as an ally or suitability as a mate. Collectively, the group can more easily detect members who take advantage of others, fail to share resources, or violate the group's norms in other ways. The exchange of social information ensures that even those who do not witness bad behavior directly will soon learn about it, and the offending party can then be punished through ridicule or ostracism. In surviving hunter-gatherer societies, such banter seems to be used to maintain equality within nomadic bands by suppressing internal competition and encouraging consensus. A group member who tries to assert his dominance or make an unreasonable claim on food or other resources may be gently teased or mocked to indicate that his peers think he is getting too big for his boots. Like grooming, gossip serves as a vital social glue.

Dunbar has gone so far as to suggest that the exchange of social information, rather than the need to pass on information about food sources or to coordinate hunting, was the driving force behind the development of language, because using language makes maintaining social bonds much more efficient—which in turn allows for larger (and safer) groups. He argues that "the most plausible starting point for the evolution of language is as a bonding device based on the exchange of social information concerning relationships within the social network." And whether or not it really did drive the development of language, the exchange of social information does seem to be the main thing for which language is used: it accounts for around two thirds of spontaneous human conversation, according to observational studies. Telling tales about others is often seen as a disreputable activity that wastes time and spreads lies or half-truths. But sharing information with other people within our social networks is, it would appear, a central part of being human.

## THE DAWN OF SHARED MEDIA

In modern societies the exchange of social information is not limited to people who are physically present. Just as gossip is grooming at a distance, various forms of media make possible gossip at a distance, by capturing information so that it can be sent across time and space. We can exchange social information with our friends on the phone, by letter, or online. Newspapers, magazines, television, and digital media also allow us to track (or, at least, provide the illusion of tracking) the shifting social relationships between film stars, politicians, business leaders, and other celebrities, even though we have never met them in person. This is simply the desire to pass around information about common acquaintances' control of resources, sexual activity, alliances, and disputes operating at a societal level. The innate human propensity to share such information, it seems, will take advantage of any available means to do so. For most of the one hundred thousand years or so since the dawn of language, however, the only available means to convey specific items of news was speech. A new way to exchange information with others only emerged five thousand years ago, with the invention of writing.

The development of writing was pioneered not by gossips, storytellers, or poets, but by accountants. The earliest writing system has its roots in the Neolithic period, when humans first began to switch from a nomadic existence of hunting and gathering to a settled lifestyle based on agriculture. This transition began around 9500 B.C. in a region known as the Fertile Crescent, which stretches from modern-day Egypt, up to southeastern Turkey, and down again to the border between Iraq and Iran. Writing seems to have evolved in this region from the custom of using small clay tokens to account for transactions involving agricultural goods such as grain, sheep, and cattle. The first written documents, which come from the Mesopotamian city of Uruk and date from around 3400 B.C., record allocations of bread and beer rations, payment of taxes, and other transactions using simple symbols and marks on clay tablets.

Initially this primitive form of writing was not flexible enough to record spoken language, and was used simply for accounting purposes. The most common symbols represent bread, beer, sheep, cattle, and textiles. Each tablet was more like a record in a database than a piece of prose. But over the next five hundred years, writing evolved to become a more powerful and expressive medium. Fiddly pictograms scratched into clay soon gave way to more abstract symbols, or ideograms, consisting of multiple wedge-shaped ("cuneiform") impressions made using a stylus. These ideograms no longer bore much resemblance, if any, to the pictograms and tokens from which they were derived, but this approach had the advantage of making writing much faster. Another innovation, around 3100 B.C., was the use of ideograms to stand for particular sounds, probably as a result of the need to write people's names. The result was the first general-purpose form of writing.

By this time writing had emerged in Egypt, too. The Egyptian hieroglyphic system was also based on pictograms that sometimes represented things and could also represent sounds. But unlike in Mesopotamia, where the gradual progress from tokens to pictograms to cuneiform writing is clearly visible, writing appears almost overnight in Egypt, which suggests that the idea might have been imported from Mesopotamia. In any case, by 2600 B.C. both cuneiform and hieroglyphic writing systems had become flexible enough to record abstract ideas such as hymns, religious texts, and collections of advice known as wisdom literature. For the first time, it was also possible to share information with someone else by sending a message in physical form, rather than relying on a messenger to deliver the information in spoken form.

The earliest known letters date from this period. In Egypt they were written in ink on papyrus, a paper-like material made from the pith of the papyrus plant, or on pieces of broken pottery called *ostraca*. In Mesopotamia letters were written in cuneiform on small clay tablets that fit neatly into the palm of the hand. Letters almost always fit onto a single tablet, which imposed a limit on the length of the message. Most of the letters that survive from the third millennium B.C. are formal correspondence between kings and officials, rather than informal personal messages. And they are not letters in the modern sense, namely documents written by the sender to be read by the recipient. Instead they are a means of recording and transmitting spoken words from the sender to the recipient, using scribes to turn speech into written marks at one end, and then back into speech at the other. Rather than beginning with a salutation, such as "Dear so-and-so" or "To so-and-so," Mesopotamian letters from this period begin with direct instructions to the scribe reading out the letter: "Say to so-and-so."

The ability to read and write was limited to a tiny fraction of the population for the first fifteen centuries after the invention of writing, for a number of reasons. Acquiring literacy required extensive training, which was time-consuming and expensive, and therefore available to only a small subset of the elite. And the scribal class that emerged wanted to protect its privileged position as an information priesthood. It had no interest in making literacy easier to acquire and thus more widely available. Egyptian scribal-training texts emphasized the superiority of being a scribe over all other career choices, with titles like "Do Not Be Soldier, Priest or Baker," "Do Not Be a Husbandman," and "Do Not Be a Charioteer." This last text begins: "Set thine heart on being a scribe, that thou mayest direct the whole earth." Literacy was power. This probably explains why the innovation of alphabetic writing—in which a small group of symbols, standing for particular sounds, can be used to write any word—was not embraced, even though it was used in both cuneiform and hieroglyphic writing to render names and foreign words from early in the third millennium B.C. In theory, the scribes could then have abandoned the hundreds of specialist symbols they were required to learn and simply used alphabetic script instead. But this would have made writing much easier for others to learn, and thus threatened their special status.

In both Egyptian and Mesopotamian cultures, literacy was therefore largely limited to scribes and court officials. Members of the royal family seem to have had some scribal training, but they were rarely as fluent in reading and writing as were dedicated scribes and personal secretaries. Rulers would have dictated outgoing documents and had incoming ones read aloud to them. Assessing the extent of literacy in the ancient world is a difficult business, however. For one thing, there is no simple definition of literacy: should it be defined as being able to write your own name, for example, or being able to write a short statement? As with computer literacy today, there is no clear line between literacy and illiteracy. Instead, people fall along a spectrum of abilities. But the use of writing for personal communications remained impractical for most people, for whom the easiest way to send a message was to ask someone else (such as a friend) to deliver it in person. An unusual glimpse into everyday life between the sixteenth and eleventh centuries B.C. is provided by letters and other documents from Deir el-Medina, the village of the workers who built the tombs in Egypt's Valley of the Kings. The concentration of skilled workers meant that the literacy rate was unusually high: friends and family members wrote to each other both on papyrus and on flakes of stone, and their messages were delivered by friends, children, and, in some cases, the local policeman. But this is a highly unusual case. Literacy was mostly limited to a stratum within the ruling class, and was primarily used within this social group for record keeping and other administrative purposes.

Literacy became more widespread in Ancient Greece, where the earliest true alphabet (with symbols for both consonant and vowel sounds) emerged in the early eighth century B.C. Purely alphabetic writing had its origins in the Canaanite script devised around 1800 B.C. by the trading peoples of the Levant. They were exposed to both cuneiform writing (which had by this time been adapted to many languages in the region) and the hieroglyphic approach used in Egypt, and they devised an alphabet that consisted solely of consonants. It was spread by the Phoenicians, a seafaring people who established trading posts and independent city-states along the south and east sides of the Mediterranean basin. The Greeks made the crucial addition of five signs for vowels to this alphabet, which made learning to read and write much easier. Greek mercenaries left graffiti at Abu Simbel in Egypt in 593 B.C., for example, which is evidence of the wider spread of literacy in Greek society. Around this time, Greek writing also appears on coins and on black-figure vases, indicating its commercial and domestic use by ordinary people.

Hipparchus, a member of the ruling class in Athens in the sixth century B.C., had stones put up along the roads leading out of the city, labeled "This is a monument of Hipparchus," with pithy sayings underneath such as "Do not deceive a friend." This sort of self-promotion would not have been worthwhile unless a reasonable proportion of the population was literate. The introduction of the practice of ostracism in Athens around 510 B.C. also implies quite widespread literacy, at least among male citizens. The word *ostracism* is derived from the ostraca on which at least six thousand voters (or around 20 percent of male citizens) were required to write the name of anyone they wished to expel from the city for ten years. The pottery shards were then counted up, and if the threshold was reached, the person in question was banished. This practice was used to defuse political struggles by banishing one of the participants.

With their discursive political and intellectual culture and an unusually high rate of literacy, the Greeks had the opportunity to create the first social-media culture, based on the exchange of written rather than spoken information. The prefaces of mathematical works by Archimedes and Apollonius of Perga show that both men sent copies to mathematicians in other parts of Greece. But there is no evidence for a wider culture of copying and sharing of documents; Greek culture, it seems, never quite shook off its skepticism toward writing. Writing was seen as a threat to the supremacy of the spoken word, which was central to Greek culture. Political, legal, and philosophical arguments took place through face-to-face dialogues and debates. There was no need for a scribal bureaucracy, because each city-state was independent and could conduct its affairs through public meetings of its citizens. Indeed, Aristotle's definition of the ideal city specified that its population should not be so large that a single speaker could not address all of its citizens. Rhetoric, which originally meant oratory, or the art of speaking, was venerated as an important skill. And the long traditions of Greek poetry and drama were based on the spoken rather than the written word.

The Greek case against writing was summarized by Plato in the fourth century B.C. in two written works: the *Phaedrus* and the so-called Seventh Letter. The *Phaedrus* takes the form of a dialogue between Plato's teacher, Socrates, and an interlocutor called Phaedrus. Socrates complains that writing undermines the need to remember things and weakens the mind, creating "forgetfulness in the learners' souls, because they will not use their memories; they will trust to the external written characters and not remember of themselves." Socrates also points out that written texts cannot respond to queries ("if you ask them a question they maintain a solemn silence") and are subject to misunderstanding or distortion ("if they are maltreated or abused, they have no parent to protect them; and they cannot protect or defend themselves"). Socrates concedes that written texts have their uses "as memorials to be treasured against the forgetfulness of old age," but he is far more worried about their shortcomings. People who rely on written documents, he fears, will be "hearers of many things and will have learned nothing; they will appear to be omniscient and will generally know nothing."

Plato advanced a similar argument, that a written text is no substitute for a spoken dialogue, in the Seventh Letter. "Every man of worth, when dealing with matters of worth, will be far from exposing them to ill feeling and misunderstanding among men by committing them to writing," he declared. He was criticizing his pupil Dionysius, the ruler of Syracuse, who had written a philosophical treatise. Mere written words, Plato argued, cannot convey the flashes of insight through which philosophical truths come to be understood. Instead, like Socrates, Plato preferred the dialectic approach of arriving at truth through dialogue: "In the course of scrutiny and kindly testing by men who proceed by question and answer without ill will, with a sudden flash there shines forth understanding about every problem." His pupil Aristotle believed that "spoken words are the symbols of mental experience, and written words are the symbols of spoken words," another expression of the Greek view of the superiority of speech over writing.

These objections are assumed to reflect disquiet in Greek society at the mental and cultural changes associated with the spread of writing and literacy. Plato, speaking both as himself and through the character of Socrates, did not denounce writing outright; but he gave voice to concerns that intellectual life would suffer as a result of the shift away from a purely oral culture. There is a clear parallel with the similar concerns that have been expressed in recent years with regard to the benefits and drawbacks of digital media: on the one hand it is now much easier to access information quickly, but why bother to remember anything if you can simply Google it? In both cases the new technologies are considered to be unnecessary crutches for the lazy. In fact, both extend the human mind by freeing it from the need to remember absolutely everything perfectly, and allowing half-remembered items to be retrieved from written sources when needed.

The great irony is that Plato's arguments survive to this day because he wrote them down. But he did so in a transitional form: as written versions of spoken dialogues. Similarly, in Plato's time people began to enjoy reading plays as well as watching performances of them. Once again, writing was being used as a proxy for the spoken word. Such hybrid spoken-written forms highlight the way in which Greek culture straddled the transition from a purely oral culture to a hybrid oral-literary one. But any remaining worries about the adoption of writing had long since been forgotten by the time Greek culture had been overshadowed and absorbed by Rome. Literacy was moderately widespread, and educated Romans were confident users of the written word, in both formal and informal contexts. The stage was set for the emergence of the first social-media ecosystem.

# CHAPTER 2

# THE ROMAN MEDIA: THE FIRST SOCIAL-MEDIA ECOSYSTEM

*You say my letter has been widely published: well, I don't care. Indeed, I myself allowed several people to take a copy of it.*
—Marcus Tullius Cicero, *Ad. Att. 8.9*

## A PEOPLE-POWERED NETWORK

BY THE FIRST century B.C. Rome had become the undisputed master of the Mediterranean basin. After defeating Carthage, its old enemy to the south, and completing the conquest of Greece to the east, Rome had expanded its territories to include most of modern-day Spain, France, and Turkey and a large swath of the north coast of Africa. But as its influence grew its political system came under increasing strain. Rome's system of government remained essentially that of a small city-state, with power concentrated in the hands of the small group of intermarried families that made up its elite political class. Conflicts within the ruling class and between it and the wider populace led to a succession of plots, rebellions, and civil wars, interspersed with attempts at political reform. During this period of upheaval the fortunes of Rome's vast territories depended to an inordinate extent on the personal ties between members of the Roman elite. Social gossip and political news were intertwined. Amid endless alliances, conspiracies, and intrigues, maintaining a wide network of contacts, to keep up with the latest developments and stay on the right side of these constant machinations, could be a matter of life or death.

Within the city of Rome such information was mostly exchanged in person, at the Forum (the center of political and commercial activity) and at the elaborate dinner parties, or *convivia*, of which the Romans were so fond. But those outside the city, whether posted to remote provinces or relaxing at their country villas, could also participate in the exchange of information by writing letters. The fortunes of people in Rome often depended on trade or war at the fringes of the Roman world, and those on the periphery needed to keep up with the latest power struggles in the capital. For the members of Rome's ruling class, written correspondence was both an important means of distributing information and a way to define and maintain relationships with others. The Roman elite were well educated and highly literate; transport links were fast and reliable, at least by the standards of the ancient world; and the ready availability of scribes and messengers, many of them slaves, made copying and delivering messages quick and affordable, at least for the elite. Rex Winsbury, a historian of the Roman book, has called slavery "the enabling infrastructure of Roman literature," but another way to put it is that slaves were the Roman equivalent of broadband. For the first time in history news and gossip, both personal and political, began to circulate in large quantities in written form.

The messages exchanged were sometimes formal, but could also be familiar and conversational in tone and often contained colloquial terms, in-jokes, puns, and abbreviations. One common abbreviation used in Roman letters was SPD, which was short for *salutem plurimam dicit*, or "sends many greetings." This served as a greeting at the beginning of a letter, to indicate the sender and the receiver, as in "Marcus Sexto SPD" ("Marcus sends many greetings to Sextus"). Another popular acronym was SVBEEV, which was short for *si vales, bene est, ego valeo* ("if you are well, that is good, I am well"). Such abbreviations saved space and time, just as acronyms (BTW, AFAIK, IANAL) do today in Internet posts and text messages.

The variation in tone in Roman correspondence can be seen in the nearly nine hundred surviving letters of Marcus Tullius Cicero, whose correspondence is the best-preserved of the period. Some of his letters were written while traveling, or between the courses of a meal; others were formally composed with the aid of a scribe. When writing to close friends, Cicero often threw in Greek words, which both reflected his admiration of Greek culture and invoked the fraternity of the educated elite. Cicero and his web of contemporaries became so used to exchanging information by letter, with messengers coming and going throughout the day, that they considered it an extension of spoken conversation. As Cicero wrote to one of his correspondents, "we will avail ourselves of the boon of letters, and thus secure almost the same objects in our separation as if we were together." Similarly, his contemporary and rival Julius Caesar was said to have engaged in "a form of conversation with his friends via letter-writing; he could not wait to see them face to face on urgent matters, because of the excess of his duties and the great size of the city of Rome."

For wealthy Romans, the distinction between letter writing and conversation was further blurred by the custom of dictating outgoing letters to scribes, and having incoming letters read aloud to them. Efficient, trustworthy scribes were highly sought after, because they could greatly increase the amount of information a single person could handle each day. Julius Caesar was said to have been able to dictate two or more letters at once, to different scribes. Cicero had the same secretary for much of his career: Tiro, a slave who continued to work for Cicero after being granted his freedom. Tiro's unusually high throughput is said to have been the result of his invention of a form of shorthand that enabled Cicero to dictate letters and speeches quickly and handle a far higher volume of correspondence than would otherwise have been possible.

Messages sent over short distances, for which a quick reply was expected, were written with a stylus on wax tablets mounted in wooden frames that folded together like a book. To modern eyes these tablets, with their flat writing surfaces surrounded by a wooden frame, look strikingly similar to tablet computers. The recipient's response could be scratched onto the same tablet, and the messenger who had delivered it would then take it straight back to the sender. The tablets could be erased and reused by smoothing the colored wax with the flat end of a stylus. Within the city, this was a handy way to send a quick query to someone and get a reply within an hour or two. (In a letter to his friend Lepta, who had inquired about a point of law, Cicero mentions that he had fired off a message like this to another friend in order to get a definitive answer.) Letters sent over longer distances were written on papyrus, which was more expensive but lighter and therefore more suitable for transport. A single sheet of papryus typically measured about six inches wide by ten inches tall, which was enough for a short

Image of a Roman wax tablet, which looks very like an iPad.
*Romano-Germanic Museum, Cologne*

letter; but if necessary several sheets could be pasted together side by side to provide a wider writing area. Messages were written with a reed pen dipped in ink derived from cuttlefish, or from a mixture of soot, gum, and water. The words were arranged in columns, two to four inches wide, running left to right across the width of the papyrus, only one surface of which was used for writing. The finished document would then be rolled up and secured with a thread and a wax seal.

In Cicero's time, wealthy households and public officials had their letters carried by messengers known as *tabellarii*. "You have most unreasonable letter-carriers," Cicero complained to his friend Cassius. "They come to me with their traveling caps on, declaring that their company is waiting for them at the city gate." Evidently tabellarii followed regular routes for their masters, collecting and delivering letters as they went. But for ordinary people there were less extravagant ways to have letters delivered. It was normal practice to ask friends who were traveling to carry letters, if not all the way to the recipient, then at least in the right direction so that they could be passed, in turn, to other friends who might carry them closer to their destination. Cicero was friendly with the directors of the financial associations that levied taxes in Rome's foreign provinces, having acted on their behalf in legal cases, and their couriers sometimes carried his letters along with their own. This informal system of message delivery, in which people passed on messages on behalf of their friends (or friends of friends) was not as quick or secure as having a letter delivered by

Fresco from Pompeii of a husband and wife holding the paraphernalia of writing: he holds a scroll and she holds a folding wax tablet with stylus. The message this image sends is: we are literate, and proud of it. *Walter Rawlings/Mary Evans Picture Library*

one's own *tabellarius*, but it was much cheaper. Even when a formal postal system was later established under the emperor Augustus, with an organized network of horses, carriages, and post-houses, it was only for official use. Most Romans continued to rely on informal distribution of letters via their social networks.

The ease and speed with which letters could be delivered, at least between members of the wealthy elite, allowed Cicero to maintain daily contact with some of his correspondents, such as his friend Atticus. His letters show many periods in which Cicero, staying at his villa in the country, wrote to Atticus in Rome every day and implored him to write back, even if there was not much so say. Cicero relished the social contact and was hungry for news from the capital. Sending a letter every day meant that Cicero's messenger could then bring a reply from Atticus. "Still I cannot help sending to you every day to get a letter from you," Cicero wrote to Atticus on one occasion. "Whether you have any news or not, write something," he pleaded on another. "I shall write to you nearly every day, for I prefer to send letters to no purpose rather than for you to have no messenger to give one to, if there should be anything you think I ought to know." Cicero could not bear to be out of the loop.

At the other end of the scale was Seneca, a statesman and philosopher of the first century A.D., who mocked the eagerness with which some of his fellow Romans anticipated the arrival of mail from afar. In one of his letters to his friend Lucilius, Seneca described how he watched people rushing down to the port to greet the arrival of the mail boats. "While everybody was bustling about and hurrying to the waterfront, I felt great pleasure in my laziness, because, although I was soon to receive letters from my friends, I was in no hurry to know how my affairs were progressing abroad, or what news the letters were bringing," he wrote. In Roman times, like today, it seems that some people checked their mail rather more obsessively than others.

Once a letter arrived, it would be read by its recipient, and its contents might also be read out to the recipient's family and friends, or shared with business associates. Roman letters were considered semipublic documents and were often intended for a wider audience than just the recipient, in part because securing letters in transit was so difficult in practice. In many of his letters Cicero clearly censored himself to avoid saying anything politically dangerous, because there was no telling where a letter might end up. In a few cases he explicitly asked that a letter not be read to others, but this was the exception rather than the rule. Recipients were expected to be able to judge whether it was appropriate to share particular letters more widely or not. One way to tell was whether a letter was in the hand of the sender or his scribe; particularly sensitive or intimate passages might be written by the sender directly. Such sensitive messages could be sent only via a trusted carrier.

Cicero's correspondence illustrates how letters and other documents were commonly copied and sent to others. In a letter to his friend Atticus he writes, "I sent you on the 24th of March a copy of Balbus' letter to me and of Caesar's letter to him. Then on that very day from Capua I got a letter from Quintus Pedius saying that Caesar had written to him on the 14th in the following terms." In other letters Cicero writes, "Caesar has sent me a short letter of which I subjoin a copy"; "Anthony has sent me a copy of a letter from Caesar"; "Lamia was with me after you left and brought me a letter Caesar had sent to him"; and "a copy of your letter to Lentulus, the consul, was brought to me," to give just a few examples. When writing to someone, Roman correspondents might on occasion also send copies of their letter to other friends as well, thus putting it into wider circulation. Cicero and other Roman politicians also distributed copies of noteworthy speeches in this way, so that people who had not heard a speech in person could read it (and take a copy of it) instead. After the suicide of Cato the Younger, one of Julius Caesar's most vociferous opponents, Cicero and others wrote tributes to him that circulated in written form. Caesar responded by releasing an essay of his own, entitled *Anticato*. The copying and sharing of documents meant that letters, speeches, and essays could quickly ripple through the Roman elite to find a wide readership and generate further discussion.

Roman authors varied their writing styles depending on the audience they were addressing, assuming a formal tone in documents that they expected to be widely circulated and a more casual style in personal correspondence. "I write in one style what I expect that the persons addressed only, in another what I expect that many will read," Cicero once explained. Writing in the spring of 49 B.C., he expressed his satisfaction that one of his letters had been broadly distributed: "You say my letter has been widely published: well, I don't care. Indeed, I myself allowed several people to take a copy of it." Such copying was possible because Cicero often kept copies of outgoing letters, so that they could be referred to and copied for other people when necessary. On this particular occasion, with the Roman republic sliding toward a civil war between Caesar and his rival Pompey, Cicero wanted his letter to be widely shared because "I wished my sentiments as to keeping the peace to be put on record." He did so by giving copies of his letter to others, knowing that they would share them, in turn, with their own social circles.

### SHARING THE NEWS

"I feel sure that a gazette of transactions in the city is being sent to you. If I had not thought so I would have written an account of them myself," Cicero wrote to his friend Quintus Cornificius, the governor of Rome's African territories, in 44 B.C. This gazette, known as the *acta,* was the nearest thing the Romans had to a newspaper. But it was rather an unusual one. Although it was published daily, only one official copy of each issue was produced. The state took care of the initial publication of the acta in the Forum, but made no attempt to copy or distribute it. That was left instead to the acta's readers. The earliest ancestor of the newspaper relied on informal distribution via social networks to reach a wide audience.

The founder of this pioneering publication was none other than Julius Caesar. In 59 B.C. he was elected consul, the highest office of the Roman republic, to which two men were elected each year, taking turns to act as consul in alternating months. Each consul had the power of veto over the other, a measure intended to prevent abuse of power, at least in theory. In practice, Caesar's powerful political connections enabled him to sideline his co-consul Marcus Calpurnius Bibulus almost immediately, giving him free rein for the remainder of his term in office, the first of four occasions on which he would serve as consul. According to the Roman historian Suetonius, Caesar's first act after his election was to order that "the proceedings both of the senate and of the people should day by day be compiled and published." The resulting gazette was called the *acta diurna populi Romani*—literally, the "daily acts of the people of Rome"—though the name is often abbreviated to the acta diurna or simply the acta. It continued in some form until the third century A.D. The word *diurna,* meaning diurnal or daily, is the root of the English words "journal" and "journalism."

Establishing the acta diurna was a political move. Caesar was a populist who felt that too much power was concentrated in the hands of the optimates, or aristocrats, in the senate. He had been elected after promising that he would redistribute state-owned lands to the people, a policy the optimates fiercely opposed. Under the Roman system of governance, the assemblies of the common people took place in public, in the Forum. The senate, however, met behind closed doors and released the details of debates, speeches, or votes only when it wanted to. Mandating the publication of a brief summary of its proceedings each day was therefore a handy way for Caesar to highlight the aristocratic senators' opposition to his populist policies and subtly undermine the senate's mystique and authority. Caesar's aim was not to make Roman politics more open and democratic, but to undermine the senate and further his own ambition to become Rome's absolute ruler. Only by concentrating authority in the hands of one man, he believed, could Rome's chaotic politics be tamed. This brought him into conflict with the optimates, who wanted to defend the power of the senate, and with Cicero, who saw his role as protecting the republic's institutions and traditions from an ambitious tyrant.

The material in the acta came from a variety of sources. It was customary for copies of proposed legislation, noteworthy speeches, and proclamations from the people's assemblies to be posted in the Forum so that anyone who wanted to could consult and copy them. More formal records were also kept of the proceedings of the senate. All this was summarized for inclusion in the acta diurna so that it recorded the proceedings of both the people and the senate, as Caesar had specified. A further source of information was the white noticeboard, or *album*, that stood at the east end of the Forum, near the residence of the Pontifex Maximus, the chief priest of the Roman state religion. Caesar had orchestrated his own election to this important post in 63 B.C., and it gave him responsibility for consecrating temples, regulating the calendar, overseeing aristocratic marriages, ruling on wills and adoptions, and regulating the public morals, among other things. The album was an official noticeboard that displayed the names of each year's ruling consuls and other state officials, along with the dates of holidays, religious festivals, and eclipses and announcements of important births, deaths, marriages, and divorces—all of which could now also appear in the acta.

The resulting bulletin was posted on wooden boards in the Forum each day. Initially the acta focused on the proceedings of government, providing accounts of speeches and the names of senators who supported or opposed particular bills. This suited Caesar's purposes, though he also made more direct political use of the acta. When, at the height of his power in 44 B.C., he made great show of refusing the title of king, for example, he made sure that the acta reported this act of supposed humility. (Caesar had, by this time, been declared "dictator for life," so he was already king in all but name.) Within a few years of its founding the scope of the acta had been widened to include nonpolitical items, such as gifts and bequests, funerals, and unusual and noteworthy occurrences. Even these apparently trivial matters could have political significance. Divorces were a subject of much interest in Roman politics, for example, because marriage was commonly used to reinforce political alliances or acquire the wealth to fund political careers. Divorces therefore signaled political opportunity. Seneca wryly noted that "there are no acta without divorces."

The acta did not go in for big headlines and sensationalist prose. Instead it related events in a brief and telegraphic style, rather like a modern newswire, judging by the parody provided by the dramatist Petronius in his *Satyricon*, a work of fiction from the first century A.D. In one passage Trimalchio, a former slave who has become fabulously rich, holds a lavish but vulgar feast. A clerk appears and informs the assembled guests of the latest developments on Trimalchio's vast estates, with a bulletin written in the style of the acta: "On the seventh of the Kalends of Sextilis (August), on Trimalchio's estates near Cumae, were born thirty boys and forty girls. Five hundred bushels of wheat were taken from the threshing floors and stored in the granaries. Five hundred oxen were put to yoke. The slave Mithridates was crucified on the same date for blaspheming the genius of our master, Gaius [Trimalchio]. On said date ten million sesterces were returned to the vaults because no sound investment could be found. On said date, a fire broke out in the gardens at Pompeii, said fire originating in the house of Nasta, the bailiff."

Alongside political reports and official statistics and announcements there were evidently items of trivia. Pliny the Elder, a Roman naturalist and statesman who lived in the first century A.D., drew upon archived copies of the acta when compiling accounts of unusual natural phenomena for his *Naturalis Historia*, the world's first encyclopedia. He tells, for example, of a shower of rocks that fell on the Forum during a speech by Milo in 50 B.C.; of a grief-stricken fan of a charioteer of the red faction who threw himself onto the charioteer's funeral pyre; of a man who came to sacrifice at the Temple of Jupiter in Rome with his eight children, twenty-eight grandsons, eight granddaughters, and nineteen great-grandchildren; and of a phoenix that was brought to Rome and displayed in the Forum, "a fact attested by the records, although nobody would doubt that this phoenix was a fabrication." Another report cited by Pliny concerns an unusually loyal dog that refused to abandon the corpse of its master, who had been executed: "When his corpse had been thrown into the Tiber it swam to it and tried to keep it afloat, a great crowd streaming out to view the animal's loyalty."

Despite (or perhaps because of) the inclusion of such items, the acta was avidly read within Rome and across the Roman world. The historian Tacitus, writing in the first century A.D., explains that news published in the acta became known "throughout the provinces and throughout the army." In his letters, Cicero routinely assumes that his friends in far-flung provinces have access to the acta, and he refers to receiving his own copies while traveling ("I have the city gazette up to March 15"; "I await at Thessalonica the gazette of the proceedings of August 1"). These copies were all produced and distributed by hand. The Romans had the technology needed to make a moveable-type printing press: in one of his philosophical treatises, Cicero himself even mentioned, in passing, the possibility of making up words from individual letters ("Why not believe, also, that by throwing together indiscriminately, innumerable forms of letters of the alphabet, either in gold or any other substance, one can print on the ground with these letters, the annals of Ennius?"). But slavery meant that human copyists were cheap, so there was little incentive for anyone to devise a press.

Within Rome, the wealthy could send a scribe to the Forum with a stylus and a wax tablet to jot down items of interest from the acta, along with any other documents of interest or gossip available at the Forum. Those who wanted the entire contents of the acta could send a scribe to copy it all down. But it soon became apparent to some enterprising Romans that there was money to be made from noting down the contents of the acta each day, having multiple copies written out by scribes, and delivering those copies to paying customers within the city. Juvenal, in his *Satires*, describes an aristocratic woman reading the "lengthy journal" (presumably a copy of the acta) at home while having her makeup applied and tormenting her slaves. These copies could, in turn, be copied by household scribes and sent to friends outside the city. Cicero's letters show that he and his contemporaries relied on copies or excerpts of the acta, obtained in this way and sent to them by their friends in bundles with letters, to keep up with events when they were away from Rome. A century and a half later, Pliny the Younger was still getting his news in this manner when visiting his country estate in Tuscany. "Don't drop your habit of sending me the city gazette while I am rusticating in this way," he wrote to his friend Pompeius Falco in Rome.

One advantage of getting friends to pass on the news was that they could highlight items of interest and add their own comments or background information in the covering letters they sent along with the copied acta. The combination of personal letters and impersonal news was more valuable than either in isolation, because each provided additional context for the other. And then, as now, one was far more likely to pay attention to something if a friend said it was important or expressed an opinion about it. Accordingly, when Cicero was made governor of Cilicia, a remote Roman province, he asked his friend and protégé Marcus Caelius Rufus to keep him informed of goings-on in Rome. Caelius was a well-connected young man who would, Cicero hoped, provide valuable commentary on the political news and flesh out the bare-bones reporting of the acta. But Caelius misunderstood the request and assumed that Cicero simply wanted as much information as possible to be sent from the city. Caelius wrote:

> As you were leaving me, I promised to write you a very careful and full account of all that happened in the city. Well, I have been at some pains to get hold of a man who would report every detail—so minutely, indeed, that I fear you will regard his efforts in that line as a mere excess of loquacity . . . but the packet I send you herewith itself explains my conduct. It would require I don't know how much leisure, not only to write out all this, but even to cast an eye over it. Decrees of the senate, edicts, gossip, rumors—they are all there. If you are not altogether pleased with this sample, be sure you let me know, so that I may not exhaust your patience and my purse at the same time.

Caelius had employed a scribe called Chrestus to supply a copy of each day's acta, along with a compilation of other announcements, proclamations, and gossip from the Forum. We can picture Cicero unwrapping one of the scrolls of news he had received from Caelius, settling down to read—and frowning in disappointment. "Well! Do you really think that this is what I commissioned you to do, to send me reports of the gladiatorial pairs, the adjournment of trials, Chrestus' compilation, and such tittle-tattle as nobody would have the impertinence to repeat to me when I am at Rome?" Cicero wrote back to Caelius. He explained that what he really wanted was for Caelius to use his judgment to filter and comment on the news, providing a perceptive analysis of political events, not the gladiator results. "That is why I do not look to you for anything about the past or present, but, as may be expected of a man who sees so far ahead into the future, about what is likely to happen." A somewhat chastened Caelius wrote back to Cicero to justify his decision to send him so much information. "There is much which you must skip, especially the detailed accounts of the games and funerals, and all the rest of the tittle-tattle. But the great part is useful. In fact I had rather err in the direction of telling you what you don't desire to know, than that of passing over anything that is essential."

Infuriatingly for historians, not a single copy of the acta has survived. So we are unable to see how it reported the events of the most tumultuous period of Roman history: the transition from the republic to the empire. Caesar's assassination by his political opponents in the senate in March 44 B.C., the murder of Cicero at the hands of Caesar's sympathizers, and the ensuing civil war that eventually resulted in Caesar's adopted son Octavian emerging as the first Roman emperor, Augustus, would all have been covered in the acta. But unlike the letters and books of Cicero, Caesar, and their contemporaries, the columns of the acta were not thought to be worth preserving by ancient copyists, and all have perished.

The Roman writer Virgil told of Fama, the goddess of rumor, who took the form of a dreadful monster with large wings and an ear, an eye, and a mouth at the base of every feather. "In the day she posts herself on the summits of the highest buildings to see all; in the night she perambulates the firmament to relate all; she never takes repose, as assiduous in spreading the false as in distributing the true." The Romans were certainly used to hearing news from afar. In reality, however, it was not a perambulating goddess who distributed it so widely. It was scribes, messengers, and travelers bearing wax tablets and papyrus rolls who powered the Roman information system, copying and carrying messages from one friend to another.

### GENERATING SOME BUZZ

Just as news and gossip spread in Roman times through sharing along social networks, the same was true of books. The Roman literary world had no publishers, no copyright, and very few booksellers. Instead, books circulated from one reader to the next through recommendation and copying. Authors did not make any money from their books and wanted as many people as possible to copy them. The best way to do this was to get copies of a book into the personal libraries of influential intellectuals. A well stocked library would attract a steady stream of visitors, who might read and recopy the book and so help it reach a wide readership. Only the most popular books were in such demand that booksellers would have copies made for sale in their stalls; sometimes buying a book was quicker and more convenient than having scribes transcribe a friend's edition. For a book to succeed, its author had to exploit social connections, win endorsements from influential tastemakers, and generate buzz in the right circles.

Roman books, like letters, were written on papyrus, though on rather longer scrolls known as book-rolls. Each was as much as ten meters (thirty-three feet) long when fully unrolled. At each end of the scroll was a wooden rod, around which the papyrus could be rolled or unrolled with the aid of knobs on the tips of the rods. Reading a book involved turning these knobs to roll the papyrus from the right-hand rod to the left-hand one. As with letters, the text was written vertically in columns a few inches wide, though the text of a formal copy of a book was written out in neat, regular columns justified right and left. There was usually a title sheet at the beginning and end of each scroll, and a label was attached to one of the wooden rods so that books could be identified when stacked up on shelves or in book boxes. Fancier books had cylindral covers to protect each scroll.

Once an author had finished composing a book—"writing" is not quite the right word, because it typically involved repeated dictation to a scribe followed by revision and editing—a neat copy would be written out by a specialist scribe called a *librarius*. The author would usually show this to a few close friends to ask for their comments. Cicero relied heavily on advice from his friend Atticus, who was a wealthy and well-connected man with a large library and an impressive team of scribes for writing out, copying, and editing books. As well as making suggestions on the text by marking passages that needed work with red wax, Atticus made his scribes and copyists available to Cicero and helped promote his books by recommending them to his literary friends. Distributing copies of a book at this preliminary stage was frowned upon, however, because the text might not be final, and the author did not want unfinished versions of his work going into circulation. On one occasion Cicero wrote to Atticus to complain that he had allowed two people to copy parts of "De Finibus," a book of philosophy he was writing.

> Now just tell me—do you think it right, to begin with, to publish at all without an order from me? . . . And again: do you think it right to show it to anyone before Brutus, to whom, on your advice, I dedicate it? For Balbus has written to tell me that you have allowed him to take a copy of the fifth book of the de Finibus, in which, though I have not made very many alterations, yet I have made some. I shall be very much obliged to you if you will keep back the other books, so that Balbus may not have what is uncorrected.

Cicero's anger was tempered by the knowledge that Balbus was a wealthy and influential man and an associate of Caesar's, and might therefore mention the book to him. Providing a sneak preview of the book in this way might generate wider interest in it. But Cicero was also worried that the book would leak out before he had been able to give a finished copy to Marcus Brutus, to whom he planned to dedicate the work at Atticus's suggestion. By tradition, the dedicatee received the first finished copy of a new book, and the choice of dedicatee could make a huge difference to a book's fortunes. Cicero dedicated another of his philosophical works, the *Academica,* to Marcus Terentius Varro, a literary scholar and prolific author in his own right. Varro had a large private library that other scholars consulted and copied from, and by dedicating a book to him Cicero could ensure that it would be added to Varro's shelves, where it might catch the eye of these visitors. Dedicating a book to someone was also a way to win a patron, gain political favor, and highlight or reinforce the author's social connection to them. Copies of books traveled along social networks and also served to strengthen ties in those networks. In the words of historian Rex Winsbury, the sharing of prose and poetry in this way was part of the "social glue that held the upper class together."

In Cicero's day a common way to generate interest in a new work was to hold a dinner party at which excerpts would be read out by a skilled slave known as a *lector.* Reading books aloud was not easy because there were no punctuation marks or spaces between words, so rather like learning to play a piece of music, it was necessary to practice the reading of a particular text to determine the appropriate phrasing. Atticus promoted Cicero's works at his dinner parties on several occasions, starting with a speech, "Pro Ligario," which Cicero had already delivered and wished to make available in book form. "You have given my speech for Ligarius a famous start," Cicero wrote to Atticus after the party, which he had not attended. "Henceforth, whenever I write anything, I shall entrust the advertising to you." In a subsequent letter Cicero congratulated Atticus further ("You certainly had a good audience!" he wrote), noting that the text was now circulating so widely as a result of this promotional dinner that he could no longer make changes to it. A text that was only in the hands of a few people, by contrast, could still be modified. For this reason the poet Horace advised authors to wait nine years after completing a work before handing it out, to be sure that they were entirely happy with it, because "once your words are sent out you can't recall them."

Personal endorsements from influential figures could also boost a book's fortunes, which is why Cicero wrote a rather fawning letter to his colleague Quintus Cornificius, offering him a copy of one of his books. "To this book I should like you to give the support of your approval, if possible from a sincere feeling, but if not at least out of friendship." Cornificius was abroad at the time, and rather than sending a copy of the book directly, Cicero offered to make the book available to his household scribes in Rome, so that they could make a copy for him: "I will tell your people that, if they choose, they may copy it out and send it to you." Cicero hoped that Cornificius would read his book and recommend it to others.

By the end of the first century B.C. a more formal way to announce and promote a new book, called the *recitatio,* had established itself alongside the tradition of after-dinner readings of literary works. The recitatio was a launch party at which a work (or excerpts from a work) was read to an invited audience, either by the author or by a lector—but either way, the author would certainly be present. The recitatio became an institution of Roman literary life; writing in the late first century A.D., Pliny the Younger mentions that there was hardly a single day in the month of April without one. Once the reading was complete a presentation copy of the book would be handed over to the dedicatee, and other copies would also be made available to the author's friends and associates. The work was then considered to have been published, in the sense that it had been formally released by its author for reading, copying, and circulation. At that point the book was on its own and would either spread—or not, depending on whether the author had succeeded in generating sufficient buzz. One sign of a successful book was that booksellers would have copies of it made for sale to the public, something they would only do if they were sure there would be demand for it. Roman authors, then, wanted their books to be as widely copied by as many people as possible, and ideally for such copies to be put on sale, even though the author himself would not benefit financially.

Sometimes an author might not be happy that his book was being widely circulated, however. One concern was that poor transcription might render a book illegible or introduce errors that would be transmitted to all subsequent copies. (Seneca refers to "a book which we throw out because it is written in minute letters or tear up because it is full of errors.") A bigger worry, for some authors, was the circulation of unauthorized works that they had not approved for publication. Quintilian, a first-century author, complained that two of his books on rhetoric were "already in circulation under my name, though neither issued by me nor composed with that in mind." Similarly, Galen, a doctor and prolific medical writer who lived in the second century A.D., grumbled about a shorthand writer who put the transcript of one of his speeches into unauthorized circulation, and some of his students who circulated some of his writings without his permission. Worse, his authorized works were being altered and passed off by others as their own work, while medical treatises by different authors were fraudulently attributed to him. At one point Galen even overheard two people at a bookseller's stall in Rome discussing whether a book published under Galen's name really had been written by him or not. He responded by publishing a book, "On His Own Books," that included a list of his genuine works. Galen's concern was not that he would lose out financially—Roman authors, after all, made no money from their books—but that his reputation as the Roman world's greatest doctor would be damaged if substandard or fraudulent work were being passed around under his name. In truth, however, Roman authors had no control over their works once they had gone into general circulation. And for readers, the practice of distribution via sharing and repeated copying meant there was no way to check whether a book really was by a particular author, or was an accurate or authorized text, unless it had been supplied by the author himself.

## GNAEUS ALLEIUS WROTE ON YOUR WALL

Dictating and exchanging letters, sharing written news, and keeping up with the latest books required the ownership of specialist slaves and was, accordingly, limited to the Roman elite of aristocrats, generals, lawyers, state officials, and businessmen. But ordinary people came into daily contact with another form of media in which almost anyone could participate: graffiti. The walls of Roman towns and cities were covered with written messages, including advertisements, political slogans, and personal messages of all kinds. These messages, sometimes with accompanying images, were either scratched into the plastered walls of Roman buildings, painted onto them, or written with charcoal. The traditional layout for a Roman house faced inward, with rooms looking onto an internal courtyard. Facing the street was a blank wall, which provided plenty of space for graffiti. Such walls served as huge public message boards.

Very little of the plaster has survived into the modern era in most Roman ruins, having weathered away over the years. The great exceptions are Pompeii and Herculaneum, the Roman towns buried for centuries by the eruption of Mount Vesuvius in A.D. 79. They serve as time capsules in which many aspects of everyday Roman life, including the plaster walls and their graffiti, have been preserved. More than eleven thousand items of graffiti have been recorded on the walls of Pompeii, a town with a population of ten to twenty thousand. Graffiti also survive in Herculaneum and in other sites across the

Roman world, though in smaller quantities. There seems no reason to assume that the amount of graffiti in Pompeii was in any way atypical; it was probably just as ubiquitous in other Roman towns. Most of the graffiti in Pompeii are found outdoors, on the streets, but there are also examples from inside bars and brothels and, perhaps surprisingly, private houses. Writing graffiti was not regarded as defacement in the way it is today, and people commonly wrote on the walls of their own houses and those of their friends.

Historians were initially rather sniffy about the value of most of these messages and paid little attention to them. As one archaeologist put it in 1899, "the graffiti are less fertile for our knowledge of Pompeian life than might have been expected. The people with whom we should most eagerly desire to come into contact, the cultivated men and women of the ancient city, were not accustomed to scratch their names upon stucco or confide their reflections and experiences to the surface of a wall." But subsequent researchers have recognized the great potential of the graffiti to cast light on the everyday lives and social interactions of Romans of every class.

The most prominent messages, painted in large letters, were political slogans expressing support for candidates running for election to the post of *aedile*, a town official. These were either personal endorsements by prominent people ("Vesonius Primus urges the election of Gnaeus Helvius as aedile, a man worthy of public office") or endorsements by groups of their preferred candidate ("The goldsmiths unanimously urge the election of Gaius Cuspius Pansa as aedile"). Several inscriptions in Pompeii refer sarcastically to a man called Vatia in terms that imply that he is a less than worthy candidate supported only by losers: "The petty thieves request the election of Vatia as aedile"; "The whole company of late drinkers favor Vatia"; "The whole company of late risers favor Vatia." Other public notices included announcements ("Thirty pairs of gladiators provided by the magistrate Gnaeus Alleius Nigidius Maius, together with their substitutes, will fight at Pompeii on November 24, 25, and 26. There will be a hunt.") and advertisements ("To rent from the first day of next July, shops with the floors over them, fine upper chambers, and a house, in the Arnius Pollio block, owned by Gnaeus Alleius Nigidius Maius. Prospective lessees may apply to Primus, his slave.").

All of this gives a flavor of life in Pompeii, but the great merit of graffiti is that one did not have to be a magistrate like Gnaeus Alleius to add one's voice to the conversation; the walls were open to everyone. Predictably, the sexual boasts and scatological humor familiar from modern graffiti in public lavatories can also be found in Pompeii ("I screwed a lot of girls here," "Celadus the Thracian gladiator makes all the girls sigh," "Secundus defecated here."). But rather more revealing are the graffiti that simply say what people are thinking about and provide glimpses of everyday activities, rather like status updates on modern social networks.

At Nuceria, I won 8,552 denarii by gaming—fair play!
On April 19, I made bread.
On April 20, I gave a cloak to be washed. On May 7, a headband. On May 8, two tunics.
The man I am having dinner with is a barbarian.
Atimetus got me pregnant.

Witticisms and aphorisms were also popular items of graffiti ("Nobody is gallant unless he has loved," "A small problem gets larger if you ignore it"). The opening words of Virgil's *Aeneid*, "I sing of arms and the man," appear seventeen times around Pompeii, along with a punning variation of it that appears on a wall outside a launderer's shop: "I sing of launderers and their owl, not arms and the man." (Launderers were said to fear owls, which were omens of death, because people in mourning wore dark clothing that did not need cleaning very often.) Of more practical value were the messages left near inns as advice to potential customers, both positive and negative: "Phoebus the performer had a first-rate time [at this inn]"; "Once you've tried Gabinius's hotel you'll stay there"; "Would that you pay for all your tricks, innkeeper. You sell us water and keep the good wine for yourself." More practical still were the messages addressed to specific people.

Samius to Cornelius: go hang yourself!
Virgula to her friend Tertius: you are disgusting!
Sarra, you are not being very nice, leaving me all alone like this.
Lucius Istacidius, I regard as a stranger anyone who doesn't invite me to dinner.
Gaius Sabinus says a fond hello to Statius. Traveler, eat bread in Pompeii but go to Nuceria to drink. At Nuceria, the drinking is better.

There are even a few examples of dialogues, where an inscription inspired comments or responses. Two brothers, Onesimus and Secundus, sent each other messages via graffiti inscribed on walls in the house of Fabius Rufus: "Onesimus greets Secundus, his brother"; "Secundus sends very many and perpetual greetings to Onesimus"; "Lots of greetings to Secundus, lovingly." These messages seem to have been left by the brothers at the house of a mutual friend. Another dialogue was between lovers, this time on outside walls. "Secundus himself sends greetings to his Prima wherever she is. I beg you, lady, to love me." Nearby is what may be a positive response: "Prima sends very many greetings to Secundus." (Prima and Secundus were popular names that meant first-born girl and second-born son, respectively. Naming children in this way was a common Roman custom.) An exchange in which a man is ridiculed for his unrequited love was found on the wall of a tavern. The first message reads: "Successus, a weaver, loves the innkeeper's slave girl named Iris. She, however, does not love him. Still, he begs her to have pity on him. His rival wrote this. Bye, loser!" This prompted a response on the wall below it: "Envious one, why do you get in the way? Submit to a handsomer man who is being treated very wrongly and is darn good-looking." To which the reply came: "I have spoken. I have written all there is to say. You love Iris, but she does not love you. To Successus, see above. Severus."

A more high-powered discourse took place on the wall of a stairwell in the House of Maius Castricius, a well-appointed four-story residential building. Several writers contributed short fragments of poetry, in some cases quoting popular couplets that are found as graffiti elsewhere, but adding their own twists and improvements. One inscription quotes the poet Lucretius ("It is sweet on the wide sea"), perhaps referring to the house's magnificent sea views. Nearby is a poem about a gemstone: "I would like to be a gemstone [on a signet ring] for not more than an hour, so that I could give kisses to you as you sign your letters." These messages appear together in a prominent part of the house where they would be seen by anyone who entered. In this public banter, the residents of the building and their visitors seem to have been trying to outdo each other, like modern participants in a comment thread below a blog post or on a Facebook wall.

This wide range of messages, left by everyone from the residents of fancy houses to gladiators and the customers of bars and brothels, suggests that some degree of literacy was quite widespread. A literacy rate of 10 percent is sometimes given for the Roman world in the first century A.D., but even the semiliterate could write "Aufidius was here" or "Marcus loves Spendusa." And the totally illiterate could always ask someone else to write text on their behalf. This may explain the wording of one message written in the basilica, a public building next to the Forum, which reads, "Alchimus to Pyrrhus, greetings. Samanarra wrote this." The scribbling of graffiti was a world away from the elite media system, with its fancy papyrus rolls and literary soirees. A lot of what was written on walls was of no interest to anyone, as one graffito, which appears four times in Pompeii, points out: "Oh wall, I am amazed you haven't fallen down, since you bear the tedious scribblings of so many writers." But graffiti provided a vibrant, shared media environment that was open to all. As one of the thousands of messages in Pompeii puts it, "SCRIPSIT QUI VOLUIT"—"Anyone who wanted to, wrote."

## CHRISTIANITY AND SOCIAL MEDIA

The most successful users of the Roman social-media system were the followers of a charismatic Jewish preacher of the early first century. They used the sharing of media as a central part of their efforts to establish a new worldwide religion based on his teachings: Christianity. Unlike the other religions of the Greco-Roman world, early Christianity was unusual in its heavy reliance on written documents, in addition to preaching, to pass on its teachings, instruct followers, conduct debates, and resolve disputes. Starting in the mid-first century, a stream of letters and other documents began to flow between Christian churches around the shores of the Mediterranean. Of the twenty-seven works in the New Testament, twenty-one are letters (epistles), and two of the remaining six contain letters within them. In all, some nine thousand letters written by Christians survive from antiquity. Although Christians are sometimes described as "people of the book," the early church might be more accurately described as a community of letter-sharers.

The most famous of these letters are the epistles written by Paul of Tarsus, an important leader of the early church. Paul made a series of missionary journeys to establish Christian communities in Greece and Asia Minor, and as he traveled he kept in touch with the churches he had founded, and with others he planned to visit, by letter. Of the twenty-one epistles in the New Testament, fourteen are attributed to Paul (although the authorship of seven of them is contested by modern scholars). Paul's epistles were addressed to specific churches—the book of Romans is a letter to the church in Rome, for example, and Corinthians is a letter to the church in Corinth—but they were clearly intended for wider distribution. They were shared, in the first instance, by being read out to the members of the church in question. As Paul put it in the letter to the Thessalonians: "I charge you before the Lord to have this letter read to all the brothers and sisters." But he also expected recipient churches to copy his letters to other churches nearby. The second letter to the Corinthians is addressed "to the church of God in Corinth, together with all his holy people throughout Achaia" (the surrounding province); Galatians is addressed "to the churches in Galatia" (a region in central Anatolia); and Colossians instructs the recipient that "after this letter has been read to you, see that it is also read in the church of the Laodiceans and that you in turn read the letter from Laodicea." (The city of Laodicea was eleven miles from Colossae.) Paul's letters were written to be copied and shared, and they were. Readings from his letters became a part of Christian worship, and they came to be seen as scripture by the early church and were incorporated into the New Testament.

The circulation of Christian letters built on the existing customs of the Greco-Roman world. The scribes who wrote out these first Christian texts were literate members of early churches, or slaves who worked in the households of well-to-do early converts. Paul dictated his letters in the usual Roman manner; the scribe to whom he dictated the letter to the Romans is named as Tertius, for example. (At the end of the letter he sends greetings to his friends in Rome and professes his own faith: "I, Tertius, who write the letter, greet you in the Lord.") Letters were delivered by people traveling from one church to another, sometimes with the specific purpose of acting as a messenger. By the mid-first century the excellent roads and sea connections within the Roman empire had made it faster and safer for travel, and more closely interconnected, than any region of the ancient world had ever been. This encouraged more people to travel, whether for business reasons, to attend religious festivals or sporting events, or just to visit friends and family abroad. When he was not traveling, Paul based himself in cities with good communication links by land and sea, such as Corinth and Ephesus, which made it possible to send and receive letters easily.

Letters helped Paul manage his network of churches and foster the sense that they were part of a larger Christian community. His epistles repeatedly relate the particular church he is addressing both to other churches and to the Christian movement more broadly. In the first letter to the Thessalonians, for example, he likens the plight of its members, who are being persecuted by other Greeks, to that of the Christians in Judea who are also being persecuted: "for ye also have suffered like things of your own countrymen, even as they have of the Jews. He tells the Corinthians that he has asked the churches in Galatia to raise money to support the Christian community in Jerusalem, and he expects them to do the same. He also emphasizes his churches' kinship with those outside his own network, such as those founded by the apostle Peter, and the church in Jerusalem headed by James. In Christianity's early, precarious years, the feeling of being part of a wider movement was reassuring. Churches wanted to hear how other churches were faring, and whether their prayers of support for them had been answered.

It was not just Paul who wrote letters that were shared and circulated. The first letter of Peter, for example, which also appears in the New Testament, addresses Christians in Asia Minor ("the strangers dispersed through Pontus, Galatia, Cappadocia, Asia, and Bithynia") who are suffering persecution. The first letter of Clement, written right at the end of the first century, was sent by the head of the church in Rome to the church in Corinth in an effort to resolve a dispute there over the dismissal of church officials. The letter's allusions to Paul's epistles imply that copies of them were on hand in both Rome and Corinth; by this time they had probably been gathered into a collection and had become a standard Christian text. Clement's epistle shows how letters asserted the authority of church elders and were used to resolve disagreements as members of the early church argued about points of doctrine and practice. Letters written by church elders, from Paul onward, were used as evidence to support a particular theological argument. Examples include the epistles written by Dionysius of Alexandria, who wrote letters to churches across the Mediterranean world, and by Ignatius of Antioch, among others.

Indeed, the dissemination of the letters of Ignatius, who was bishop of Antioch in the late first century, shows the Christian media-sharing social network in action. Ignatius was arrested during one of the many sporadic persecutions of Christians by the Romans, who regarded the new religion with suspicion at best and outright hostility at worst. He refused to disown his Christian beliefs and was sentenced to be taken to Rome under guard for punishment, expecting to be fed to the lions in the giant arena of the Colosseum. On his way from Antioch to Rome, shackled to ten soldiers, Ignatius was greeted by delegations of Christians from many towns in Asia Minor. He wrote letters to the Christian communities in Ephesus, Magnesia, and Tralles, thanking them for their support, calling on them to be steadfast in their faith, avoid heresy, and be obedient to their bishops. He also wrote to friends in Rome to prepare them for his arrival.

At the Aegean port of Troas, while waiting for a ship, he received news that a quarrel in his home church at Antioch had been resolved. Thrilled by this news, Ignatius wrote to the nearby churches of Philadelphia and Smyrna, asking them to send letters of congratulation to Antioch. Ignatius also wrote to Polycarp, the bishop of Smyrna, asking him to inform all the churches of Asia Minor and to ask them to send letters of their own to Antioch. When passing through the Greek city of Philippi, Ignatius told the church there to send a letter to Antioch as well. The Philippians sent their letter to Polycarp in Smyrna, so that he could forward it to Antioch with other letters from churches in Asia Minor. The Philippians' letter to Polycarp also requested copies of any of Ignatius' letters that were available. Polycarp replied, "We send you, just as you requested, the letters of Ignatius, which were sent to us by him, and others, which we had with us. These are attached to this letter, and you will be able to benefit greatly from them." This indicates that Ignatius' letters were being circulated within weeks of being written, prompting one scholar to call this document-sharing system the "Holy Internet."

In some cases we even know the names of the scribes who wrote, copied, and recopied these documents, because they are named in the text or they added their names in a colophon at the end. When Polycarp was in turn arrested and executed, the church at Philomelium, near Antioch, asked the church at Smyrna to send an account of his life and death. This was duly drawn up by Gaius, drawing on the papers of Irenaeus, one of Polycarp's followers, and written out by a scribe called Euraestus. The letter sent to Philomelium, now known as "The Martyrdom of Polycarp," contained the following instructions: "When you have heard these things, send the letter to the brethren further on, so that they also may glorify the Lord." The colophon to a later copy shows that the letter was indeed copied and circulated: "This account Gaius copied from the papers of Irenaeus, a disciple of Polycarp. The same also lived with Irenaeus. And Isocrates wrote it down in Corinth from the copy of Gaius. Grace be with all men. And I, Pionius, again wrote it down from the aforementioned copy, having searched for it in obedience to a revelation of the holy Polycarp, gathering it together when it was well nigh worn out by age." Similarly, the colophon to a collection of Christian writings called the Nag Hammadi Codex VI, one of a group of fourth-century texts found in jars in Egypt in 1945, also sheds light on the practice of copying and circulation: "I have copied this one discourse . . . Indeed, very many have come to me. I have not copied them because I thought that they had come to you. Also, I hesitate to copy these for you, because perhaps they have already come to you."

The Nag Hammadi texts, like nearly all Christian texts from the second century onward, were written in codices—separate pages of papyrus or parchment, bound together along one side like a modern book—rather than being written on scrolls in the Greco-Roman style. The codex format, although favored by Christian writers, was not invented by them: the Romans and Egyptians used small notebooks in codex format, since they were smaller and easier to carry than scrolls. Roman wax tablets were often grouped together in codex format. But it seems that such notebooks and tablets were mainly used for jotting notes and recording other ephemeral information; formal documents took the form of papyrus rolls, which remained the preferred format until the mid-third century. The Roman poet Martial, writing in the late first century A.D., commended the codex format to his readers on the basis that a book could then be held in one hand and was more convenient when traveling. But his enthusiasm for the new format was not shared, with the notable exception of the emerging Christian community. By the early second century practically all Christian texts were in codex form, compared with less than 5 percent for non-Christian texts.

Quite why the Christians preferred the codex over the scroll remains unclear. One possibility is that an important early Christian document (perhaps the gospel of Mark, or a collection of Paul's epistles) was produced in codex format, and this set a trend as the document was copied and circulated. Another possibility is that early Christian writers and copyists, who would have been mostly ordinary literate people rather than specialist scribes, did not feel that they had to abide by the rules of the Greco-Roman literary elite. They were therefore happy to abandon the traditional view that codices were for notes, and that real documents should be written on scrolls. This view is also supported by the fact that Christian texts have distinctive formatting right from the start. Rather than the traditional "river of text" of a Greco-Roman document, which lacked punctuation, paragraph marks, and spaces between words, Christian documents had large letters to indicate the beginning of each paragraph. They also had marks to separate words, along with punctuation, section marks, and page numbers. All this made Christian texts much easier for ordinary people (as opposed to specialist lectors) to read aloud. So the switch from scroll to codex may be just one aspect of a wider abandonment of Greco-Roman literary customs. Once Christianity became the official religion of the Roman empire in the early fourth century A.D., the ascendancy of the codex over the scroll was assured. Of all the Greek texts that have been found in Egypt, where the dry conditions favor the survival of papyrus, the proportion written on scrolls is 98 percent for the second century, falling to 81 percent in the third century, and 26 percent and 11 percent in the fourth and fifth centuries, respectively. (In the computer era we have returned to the custom of scrolling through texts, but we now scroll up and down, rather than right to left as the Romans did.)

Along with aiding the establishment of Christianity itself, the switch from scroll to codex format was another long-standing legacy of the early church's distinctive use of media. Paul, a masterful user of the social-media system of his day, was the most influential letter-writer of antiquity, overshadowing even Cicero. In its early years, Christianity consisted of rival movements whose members disagreed over the meaning of Christ's teachings and his intended audience for them. Paul used social media to ensure that his view prevailed, cementing the establishment of the Christian church as a religion open to all, and not just to Jews. Such is his influence that his letters are still read out in Christian churches all over the world today—a striking testament to the power of documents copied and distributed along social networks.

# CHAPTER 3

## HOW LUTHER WENT VIRAL: THE ROLE OF SOCIAL MEDIA IN REVOLUTIONS (1)

*Either the Pope must abolish printing or he must seek a new world to reign over; for else as this world standeth, printing will doubtless abolish him.*
—John Foxe, *Actes and Monuments of John Foxe*, 1583

### GUTEN TAG, GUTENBERG

THE CIRCULATION OF written material in western Europe declined dramatically after the collapse of the Roman empire. During the so-called Dark Ages, between the sixth and twelfth centuries, the Christian church struggled to keep alight the flame of literacy and learning. Literacy plummeted even among the ruling classes. Texts continued to be copied for use by the church, but the process of transcribing text assumed a devotional aspect, and books took the form of elaborate illuminated manuscripts of astonishing beauty and complexity, each of which took several people months or years to produce. Papyrus from the Middle East was no longer available, so these books were written on dried animal skins, or parchment, and were produced in dedicated copying rooms in monasteries, called scriptoria. Monks would read out the text as they copied it, a form of prayer or meditation whereby the scribe absorbed the wisdom of the book. Other monks would then add elaborate illustrations in colored inks and gold leaf. The skin of a cow or sheep would only provide enough parchment for two or four pages of a large book, however, which meant that some texts required the skins of an entire herd of animals. All this made books so expensive to produce that, aside from monasteries, only noblemen or royalty could afford to own them. Such was their value that an illuminated service book, or missal, was exchanged for a vineyard by a priest in the Bavarian town of Benediktbeuern in 1074, and in 1120 the monastery of Baumberg used a missal to buy a large piece of land. Access to written knowledge became a privilege largely limited to the clergy.

Things began to change in the late eleventh century, with the rise of universities, the European rediscovery of the knowledge of the ancients (in part through exchanges with the Islamic world, which had preserved and extended it), greater trade, and rising literacy. Demand for books began to pick up. The introduction to Europe of paper, a Chinese invention, by the Arabs in the twelfth century provided a cheaper and more readily available alternative to parchment. The slow, painstaking approach of the scriptoria could turn out the small number of books needed by the church but could not meet the rising demand for nonreligious texts. As a result, universities succeeded monasteries as the main centers of book production and copying. But the expense and difficulty of copying meant that a typical university library in Europe had a very limited selection of books. In 1424, for example, the library at the University of Cambridge in England had just 122 volumes. Students would listen to a lecturer as he read from a single copy of a book, adding his own explanations as he went. Without their own copies to consult, they would rely on their own notes, or on notes taken by previous generations of students and passed from hand to hand.

The *pecia* system developed in the thirteenth century was an attempt to scale up the hand-copying of books. It involved dividing a copy of a book into sections just a few pages long, called *peciae*, and then renting them out for recopying by students or scribes, who would gradually build up a complete set of copied peciae. Allowing several copyists to copy different parts of a book at the same time in this way was much more efficient than lending out an entire book to one person for copying. It was even more efficient when students lent each other their copied peciae for recopying. (In fact, this is the architecture used by some modern peer-to-peer file-transfer systems: multiple Internet users who are downloading the same large file also exchange chunks of it with each other.) In addition to paper, another innovation, also introduced into Europe from China via the Arab world in the twelfth century, was the printing of images and small amounts of text as woodcuts, using carefully carved wooden blocks. But carving reversed text into wooden blocks was difficult and time-consuming, so it was not suitable for copying entire books.

Johann Gensfleisch zur Laden zum Gutenberg, better known as Johannes Gutenberg, was one of many people who began to look for a more efficient approach. Around 1440, after years of secretive experimentation, he managed to put together all the elements necessary for the systematic reproduction of long texts. The first element was the type itself: the individual letters, made of metal, from which words, lines, and then pages could be composed. Gutenberg, who came from a family of goldsmiths, worked out how to produce large quantities of each letter reliably and accurately using a special alloy of lead and tin and an adjustable mold that enabled him to cast letters of different widths. The second element was a carefully formulated oil-based ink that, unlike water-based ink, was viscous enough to stick to the metal type. The third element gave Gutenberg's invention its name. To apply the inked type to the paper with an even pressure, Gutenberg used a modified version of the screw press, an invention that dated back to Roman times and was used in his native Germany to make wine. Each sheet of paper or parchment (which was still used for some books in the early days of printing) was carefully brought into contact with the inked type using a folding frame, and the resulting sandwich was then slid into the press and squeezed together to transfer the ink to the paper.

Once the type for a particular page had been set, Gutenberg's system could produce hundreds of copies of it in a single day. This was more than one hundred times faster than hand-copying. As an Italian bishop observed in 1470, three men working a single press for three months could print three hundred copies of a book, whereas producing the same number of copies by hand would require a lifetime's work by three scribes. The first major work printed on Gutenberg's press, an edition of 180 copies of the Bible in Latin, sold out in advance of its release in 1455. As is often the case when one form of media supersedes another, some aspects of manuscript books survived in the first printed books: the type was based on hand-written letters, with spaces left for hand-drawn initial capitals and other illustrations to be added by hand. But printed books gradually assumed their own identity as printing took off, and new conventions emerged in typography and layout. Although Gutenberg had developed the press it was his associates Johann Fust, a goldsmith and lawyer, and Peter Schöffer, a business-savvy former scribe, who made a commercial success of it. They parted company with Gutenberg in 1455 after a legal dispute and went on to publish Bibles, Psalters, and classical texts such as the works of Cicero.

From its starting point in Mainz, knowledge of printing spread to other parts of Germany, helped along by a local dispute between archbishops that led to the sacking of Mainz in 1462, which caused many people with knowledge of printing to flee. By 1471 print workshops had been set up in several German towns and in major cities across Europe including Cologne, Basel, Rome, Venice, Paris, Nuremberg, Utrecht, Milan, Naples, and Florence—all academic and commercial centers where there was good demand for printed books. Printing reached England in 1476, when William Caxton set up the first press at Westminster Abbey, near its scriptorium. The number of presses rose sharply, and by 1500 around one thousand presses were operating in two hundred and fifty European towns and cities, and a total of perhaps ten million books had been printed. Gutenberg's printing press meant that ideas could be reproduced and circulated faster than ever before—as another German media pioneer, Martin Luther, was about to demonstrate.

## NEW POST FROM MARTIN LUTHER

Luther was a priest in the town of Wittenberg and a theologian at the local university. In 1517 he was horrified to discover that members of his congregation had been sold indulgences by a monk called Johann Tetzel. These were documents sold by the church that, according to doctrine, enabled the purchaser to avoid so-called "temporal punishment" for a particular sin and ensure a swift transition to heaven after death without having to endure punishment in purgatory. The church sold indulgences to raise the large sums required to mount military expeditions and build elaborate cathedrals. Tetzel had been granted the right to sell indulgences by Albrecht (Albert) of Mainz, the archbishop of Mainz and Magdeburg in Germany, who had in turn been granted permission by the pope, Leo X.

Luther disapproved of the whole idea of selling indulgences, considering it to be "the pious defrauding of the faithful." But he objected even more to the way Tetzel had taken liberties with the doctrine of indulgences to increase his sales. Tetzel was offering indulgences absolving people of punishment for future sins—very handy for those planning to commit adultery, say—and said that as well as buying indulgences for themselves, people could buy them for relatives who had died, freeing them from the torment of purgatory right away. Indeed, he implied, it was cruel not to spend one's life savings freeing one's ancestors from such suffering. Infuriated, Luther wrote to Archbishop Albrecht to complain about Tetzel, objecting both to Tetzel's snappy slogan—"As soon as the coin in the coffer rings, so the soul from purgatory springs"—and his brazen distortion of the doctrine of indulgences. Luther was unaware that Albrecht was keeping half of the proceeds to pay off his own debts, passing the rest back to Rome to fund the construction of Saint Peter's Basilica.

Along with his letter, Luther included a "Disputation on the Power and Efficacy of Indulgences"—a list of ninety-five theses, or propositions, written in Latin that he wished to discuss, in the academic custom of the day, in an open debate at the university. Thesis number twenty-seven, for example, asked whether Tetzel's slogan was theologically correct; that is, did payment really lead to instant release from purgatory? Number eighty-two asked why the pope did not empty purgatory right away, if it was in his power to do so; number eighty-six asked whether it was appropriate for the pope to build a lavish church in Rome using money amassed from selling indulgences to the poor. Albrecht took one look at these "Ninety-Five Theses" and sent a copy to the pope in Rome, suspecting Luther of heresy. Luther, meanwhile, announced his planned debate on indulgences in the usual way: on October 31, 1517, he posted the list of the propositions to be discussed on the door of the castle church in Wittenberg, which served as the university's noticeboard. At which point, something rather unusual happened.

Despite being written in Latin, the "Ninety-Five Theses" caused an immediate stir, initially within academic circles in Wittenberg and then farther afield. Copies began to circulate in manuscript form, and in December 1517 printed editions of the theses, in the form of pamphlets and broadsheets, appeared simultaneously in Leipzig, Nuremberg, and Basel, paid for by Luther's friends. The extent to which Luther was complicit in this publication is unclear. An enterprising printer soon produced a German translation, which could be read by a wider public and not just by Latin-speaking academics and clergymen. The "Ninety-Five Theses" spread with astonishing speed throughout the German-speaking lands as the list was copied and republished by printers in different towns. Luther's friend Friedrich Myconius later wrote that "hardly fourteen days had passed when these propositions were known throughout Germany and within four weeks almost all of Christendom was familiar with them. It almost appeared as if the angels themselves had been their messengers and brought them before the eyes of all the people. One can hardly believe how much they were talked about."

Such rapid, spontaneous distribution of a printed work in response to popular demand was an entirely new phenomenon. The unanticipated success of the "Ninety-Five Theses" revealed to Luther that there was a strong appetite for his ideas and alerted him to the way in which printed pamphlets passed from one person to another could quickly reach a wide audience. "They are printed and circulated far beyond my expectation," he wrote in March 1518 to a printer in Nuremberg who had published a German translation of the theses. But he would, he added, "have spoken far differently and more distinctly had I known what was going to happen." Luther realized that if he wanted to address a large audience about corruption in the church, then writing in scholarly Latin and having his words translated into German by others was not the best approach. Accordingly, he switched to simple, direct German for the publication later that month of his "Sermon on Indulgences and Grace," avoiding regional vocabulary to ensure that his words were intelligible from the Rhineland to Saxony. The pamphlet was an instant hit—it had to be reprinted eighteen times in 1518 alone, in editions of at least one thousand copies each time—and it demonstrated Luther's skill in exploiting the media environment of his day. Luther did not set out to split the Christian church, but that was the ultimate effect of the campaign he had started, in which the social sharing of his works played a vital role.

## LIKE, RECOMMEND, SHARE

Luther had unwittingly revealed the power of a decentralized, person-to-person media system whose participants took care of distribution, deciding collectively which messages to amplify through sharing, recommendation, and copying. It was similar to the Roman media system in many respects, but the advent of the printing press meant that copying could now be done on a much larger scale. Luther would pass the text of a new pamphlet to a friendly printer—no money changed hands—and then wait for it to ripple through the network of printing centers across Germany. Copies of the initial edition, sold for a few pfennigs (about the same price as a chicken), would first spread throughout the town where it had been printed, recommended by Luther's sympathizers to their friends. Buying an occasional pamphlet was easily within reach of a craftsman or merchant. Containing between eight and thirty-two pages, pamphlets were much cheaper than books, making them the first printed works that ordinary people could afford to buy. Owning a pamphlet, and sharing its contents with friends, signaled both literacy and endorsement of its author's ideas. Even some illiterate or semiliterate people bought Luther's pamphlets to indicate their support for his views. And if the authorities objected, a pamphlet could be easily hidden.

Traveling merchants, traders, and preachers would then carry copies to other towns, and if they sparked sufficient interest, local printers would copy them to produce their own editions, putting another thousand or so copies into circulation each time in the hope of making a quick return. As in Roman times, whether a particular item spread or not depended on the cumulative decisions made by individuals linked in social networks. But unlike in Roman times some of these individuals—the printers—wielded disproportionate influence, because they could produce a large number of copies very quickly. (In today's Internet terminology, we would call them *supernodes* in the network.) They would only reprint something, however, if they were confident that it would sell. If a printer or bookseller was repeatedly asked for a pamphlet he did not have, that was a clear sign of unmet demand. The result was that a popular pamphlet would spread quickly via this informal system of sharing, recommendation, and reprinting—without any involvement from its author.

In a modern social-media system, the popularity of a given piece of content may be inferred from the number of Likes, retweets, reblogs, +1s, or page views it generates. The equivalent metric in Luther's day was the number of separate editions of a pamphlet—in other words, the number of times it was reprinted. Of the seventy-five hundred different pamphlets that were published in German-speaking lands be-

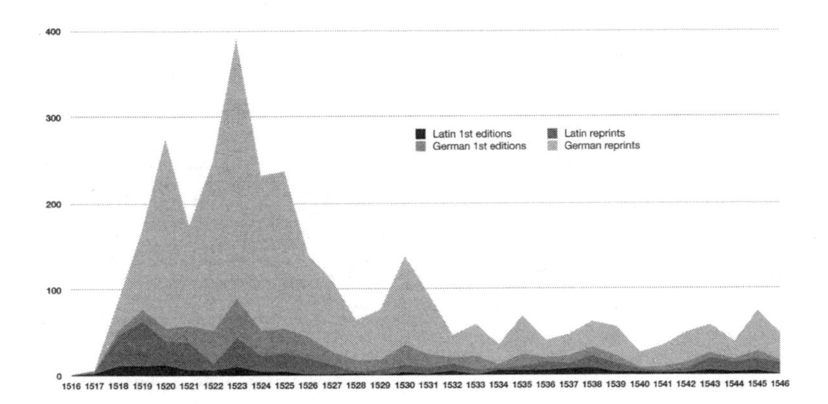


Martin Luther's traffic statistics: the number of editions of his pamphlets printed each year in German and Latin. *Based on data from Edwards,* Printing, Propaganda, and Martin Luther *(1994)*

tween 1520 and 1526, some two thousand were editions of a few dozen works by Luther. His pamphlets were by far the most sought after; a contemporary remarked that they "were not so much sold as seized." The peak year was 1523, during which nearly four hundred editions of various pamphlets by Luther were published. In all, some six million pamphlets, perhaps a third of them by Luther, were printed in the first decade of the turbulent period known today as the Reformation.

Luther was the most prolific and popular author, but there were many others on both sides of the debate. Tetzel, the indulgence-seller, was one of the first to respond to him in print, firing back with his own collection of theses. Others embraced the new pamphlet format to weigh in on the merits of Luther's arguments, both for and against. Sylvester Mazzolini, an Italian theologian, defended the pope against Luther in his "Dialogue Against the Presumptuous Theses of Martin Luther." He called Luther "a leper with a brain of brass and a nose of iron" and dismissed his arguments out of hand on the basis of papal infallibility. Luther, who was the sort of person who was incapable of letting any challenge go unanswered, took a mere two days to produce his own pamphlet in response, giving as good as he got. "I am sorry now that I despised Tetzel," he wrote. "Ridiculous as he was, he was more acute than you. You cite no scripture. You give no reasons." Johann Eck, a German theologian, wrote a letter criticizing Luther and circulated it among his friends. This prompted Andreas Karlstadt, on Luther's side, to draw up 379 theses of his own, which he subsequently expanded to 405. Eck retaliated with more theses of his own, and Karlstadt responded in kind.

As the pamphlets and letters flew back and forth, Luther's views hardened. Based on his reading of Paul's epistle to the Romans, he had reached the conclusion that faith alone was enough to merit God's forgiveness; all the paraphernalia of the church (confession, indulgences, and so forth) was irrelevant. He called into question the legitimacy of the Catholic church and challenged central items of doctrine, such as the validity of the sacraments. A passionate and mercurial character, he was capable of learned theological arguments one moment and astonishingly bawdy, colorful invective against his opponents the next. ("I confess that I have been more bitter and vehement against them than is in keeping with my Christian estate and calling," he once said of his attacks on his opponents.) When Leo X issued a papal bull threatening him with excommunication in 1520, Luther burned a copy of it in public, denounced the pope as the Antichrist, and issued even more pamphlets.

Theological arguments that would previously have taken place behind closed doors were now taking place in public, in printed form. Being able to follow and discuss the back-and-forth exchanges among Luther, his allies, and his enemies gave ordinary people across Germany a thrilling and unprecedented sense of participation in a vast, distributed debate. Arguments within their own social circles about the merits of Luther's views could now be seen as part of a far wider discourse, both spoken and printed. Many pamphlets called upon the reader to discuss their contents with others and read them aloud to the illiterate. Luther and other writers of Reformation pamphlets sometimes wrote with this in mind, presenting arguments in the form of snappy dialogues rather than prose. In one pro-Luther pamphlet from 1521, for example, a peasant is asked to adjudicate in a theological debate between Luther and one of his opponents. The views of the two parties could then be clearly distinguished (no prizes for guessing who wins the debate), and the format also emphasized that it was up to ordinary people to weigh the arguments and reach their own conclusions. By their mere existence, Luther's pamphlets embodied his belief that everyone should be allowed to participate in the debate about the reform of the church.

People read and discussed pamphlets at home with their families, in groups with their friends, and in inns and taverns. Thanks to the "marvelous, new and subtle art, the art of printing," one of Luther's contemporaries later noted, "each man became eager for knowledge, not without feeling a sense of amazement at his former blindness." Lutheran pamphlets were read out at spinning bees in Saxony and in bakeries in Tyrol. In some cases entire guilds of weavers or leather workers in particular towns declared themselves supporters of the Reformation, indicating that Luther's ideas were being propagated in the workplace. One observer remarked in 1523 that better sermons could be heard in the inns of Ulm than in its churches, and in Basel in 1524 there were complaints about people preaching from books and pamphlets in the town's taverns.

Meanwhile, printers and booksellers helped spread Reformation ideas beyond Germany to the rest of Europe. Johann Fröben, a publisher in Basel, wrote to Luther to explain that a printer from Leipzig had given him some pamphlets by Luther, "which, as they were approved by all the learned, I immediately reprinted. We have sent 600 copies to France and Spain." Fröben explained that another bookseller had taken Luther's works to Italy "to distribute among all the cities," adding that "we have sold out of all of your books except ten copies, and never remember to have sold any more quickly." Contributors to the debate that Luther had ignited ranged from the English king Henry VIII, whose treatise attacking Luther (probably written in large part by Thomas More) earned him the title "Defender of the Faith" from a grateful pope, to Hans Sachs, a shoemaker from Nuremberg who wrote a series of hugely popular songs in support of Luther.

## A MULTIMEDIA CAMPAIGN

It was not just words that traveled along the social networks of the Reformation era, but music and images, too. The news ballad, like the pamphlet, was a relatively new media format. It set a poetic and often exaggerated description of contemporary events to a familiar tune so that it could be easily learned, sung, and taught to others. News ballads were often "contrafacta" that deliberately mashed up a pious melody with secular or even profane lyrics. They were distributed in the form of printed lyric sheets, with a note to indicate which tune they should be sung to. Once learned they could spread even among the illiterate through the practice of communal singing.

Both reformers and Catholics used this new form to spread information and attack their enemies. "Ein Newes Lied Wir Heben An," Luther's first venture into the news-ballad genre, told the story of two monks who had been executed in Brussels in 1523 for refusing to recant their Lutheran beliefs. Luther's enemies denounced him as the Antichrist in song, while his supporters did the same for the pope and insulted Catholic theologians ("Goat, desist with your bleating," one of them was admonished). Luther himself is thought to have been the author of "Now We Drive Out the Pope," a parody of a folk song called "Now We Drive Out Winter," whose tune it borrowed:

*Now we drive out the pope*
*from Christ's church and God's house.*
*Therein he has reigned in a deadly fashion*
*and has seduced uncountably many souls.*

*Now move along, you damned son,*
*you Whore of Babylon.*
*You are the abomination and the Antichrist,*
*full of lies, death, and cunning.*

Woodcuts were also a useful form of propaganda. The combination of bold graphics with a smattering of text, printed as a broadsheet, could convey messages to the illiterate or semiliterate and serve as a visual aid for preachers. Luther remarked that "without images we can neither think nor understand anything." The best examples were produced by his friend Lucas Cranach. Some of his images were astonishingly graphic: "The Origins of the Pope" shows a hideous she-devil giving birth to the pope and several cardinals; "The Origin of the Monks" similarly shows three devils excreting a pile of monks. Some religious woodcuts were more elaborate than this, with complex allusions and layers of meaning that would only have been apparent to the well-educated. "Passional Christi und Antichristi," for example, was a series of images contrasting the piety of Christ with the decadence and corruption of the pope. But the images at the coarser end of the spectrum could be understood by anyone. Luther's opponents responded with woodcuts of their own: "Luther's Game of Heresy" depicts him boiling up a stew with the help of three devils, producing fumes from the pot labeled FALSEHOOD, PRIDE, ENVY and HERESY.

Amid the barrage of pamphlets, ballads, and woodcuts, public opinion was clearly moving in Luther's favor, to the dismay of the Catholic church. "Idle chatter and inappropriate books" were corrupting the people, fretted one bishop. "Daily there is a veritable downpour of Lutheran tracts in German and Latin . . . nothing is sold here except the tracts of Luther," lamented Aleander, the pope's envoy to Germany, in 1521.

The use of new media technologies to spread ideas quickly and widely posed a grave dilemma for the church. Its leaders did not want Luther's views to stand unopposed. Yet the church was reluctant to address the public with official pamphlets of its own rebutting his arguments. Arguing with Luther would be tantamount to admitting that theological matters were open to debate, and that the public had a right to evaluate competing viewpoints and make up its own mind, both of which would undermine the church's authority. And once Luther's views had been condemned as erroneous and heretical (he was formally excommunicated in 1521), providing further justification for this verdict would weaken its force by implying that it was open to argument. Some church officials also worried that responding to Luther would only serve to publicize his radical views more widely. When it came to responding to Luther, you might say, the Catholic church was damned if it did and damned if it didn't. In the modern era, companies that find themselves being criticized on social media face a similar problem. It is tempting for executives to dismiss and ignore online critics. Responding to them grants them both legitimacy and publicity. But failing to rebut critical claims may be taken by others as a tacit admission of their veracity, making matters worse.

The Catholic church faced the additional complication of the doctrine of papal infallibility, which made a direct response even more problematic. So the church decided to leave it to others, such as Tetzel and Mazzolini, to respond on its behalf. The trouble was that most of the sixty or so writers who rallied to the pope's defense did so in academic and impenetrable Latin, the traditional language of theology, rather than in German. Where Luther's works spread like wildfire, their pamphlets fizzled: Luther alone outsold all his opponents combined by a factor of five to one. One critic, Thomas Murner, even translated one of Luther's Latin pamphlets into German in order to attack its theological arguments. But people then bought Murner's pamphlet in order to read Luther's words, not Murner's criticism of them, so his effort backfired. Attempts at censorship failed, too. Printers in Leipzig were banned from publishing or selling anything by Luther or his allies, but the ban was ineffective because material printed elsewhere still flowed into the city. Eventually the city council complained to the Duke of Saxony on the printers' behalf that they faced losing "house, home, and all their livelihood" because "that which one would gladly sell, and for which there is demand, they are not allowed to have or sell." What they had was lots of Catholic pamphlets, "but what they have in overabundance is desired by no one and cannot even be given away." In a social-media system in which the audience participates in distribution, undesirable material is hard to control, and one cannot force something to spread if there is no demand for it.

Luther's enemies likened the spread of his ideas to a sickness. The papal bull threatening Luther with excommunication said its aim was "to cut off the advance of this plague and cancerous disease so it will not spread any further." Similarly, the Edict of Worms issued after his excommunication in 1521 warned that the spread of Luther's message had to be prevented, otherwise "the whole German nation, and later all other nations, will be infected by this same disorder." But it was too late: the infection had taken hold in much of Germany and beyond. To use the modern terminology, Luther's message had gone viral.

## SOCIAL MEDIA AND REVOLUTIONS

The Edict of Worms had declared Luther an outlaw, banned his works, and ordered his arrest as a heretic; anyone who gave him food or shelter was also breaking the law. But the extent of support for

Luther meant the secular authorities were reluctant to enforce the church's sentence against him. Indeed, the local ruler, Frederick III of Saxony, arranged to have Luther spirited away and taken to Wartburg Castle, where he hid for several months, firing off more letters and pamphlets and eventually shaming Archbishop Albrecht into halting the sale of indulgences in his territory. During this period Luther also translated the New Testament into German so that it could be read by ordinary people without the need to rely on the interpretation of priests. The first edition of around four thousand copies went on sale in September 1522 and sold out almost immediately.

Luther was not the first person to object to indulgences, nor to complain that the church had become corrupted by wealth and had lost touch with its original values. Similar views had been espoused by John Wycliffe, an English philosopher who lived in the fourteenth century, and by Jan Hus, a priest in Bohemia (now part of the Czech Republic) who embraced and promoted Wycliffe's ideas. Both Wycliffe and Hus were condemned by the church, and Hus was declared a heretic and burned at the stake in 1415. But where Wycliffe and Hus had had to rely on manuscript copying to spread their views, Luther had the press at his disposal, which enabled him to reach a large audience very quickly. The clamor for his writings revealed, to him and then to the public at large, that there was widespread support for his views. Luther came to regard the press as a God-given instrument that had helped him succeed where Hus had failed a century earlier.

In the early years of the Reformation, expressing support for Luther's views, whether through preaching, reading out a pamphlet, or singing a news ballad directed at the pope, was a dangerous thing to do. By stamping out isolated outbreaks of opposition swiftly, the church discouraged its opponents from speaking out and linking up. But the surge in the popularity of pamphlets, the vast majority of them in favor of reform, served as a collective signaling mechanism. In the words of Andrew Pettegree, a historian at the University of Saint Andrews, "it was the superabundance, the cascade of titles, that created the impression of an overwhelming tide, an unstoppable movement of opinion . . . Pamphlets and their purchasers had together created the impression of irresistible force." Those who went to buy a pamphlet and were told that it had sold out could tell that it was popular. They might then be more likely to express their opinions openly, knowing that support for Luther's views was widely shared. Scholars of the use of social media in modern revolutionary movements, such as those in Tunisia and Egypt, call this phenomenon "synchronization of opinion," as social media helps people realize that their dissatisfaction with the status quo, which they have been reluctant to express, is shared by others. As Luther put it in a letter to a friend after the unexpected success of the "Ninety-Five Theses," "from the rapid spread of the theses, I gather what the greater part of the nation thinks of indulgences."

Luther had given voice to the opposition to the church that had been simmering beneath the surface of European society. But the decentralized nature of his campaign meant that splits quickly emerged within the Reformation movement; revolutionaries always find it easier to agree on what they want to get rid of than what they want to replace it with. For one thing, although the reformers agreed that the authority of scripture rather than of the clergy was paramount, they disagreed on how to interpret it. Erasmus of Rotterdam, Europe's leading intellectual at the time, who thought the reformers had gone too far, scornfully observed in a letter to Luther, "You stipulate that we should not ask for or accept anything but holy scripture, but you do it in such a way as to require that we permit you to be its sole interpreter, renouncing all others." Luther's friend Andreas Karlstadt, meanwhile, wanted to do away with music and images in church, leading to a spate of riots in which religious statues and images in churches were destroyed. Other preachers took even more radical theological positions, leading to further unrest. Thomas Müntzer, another German theologian, went farthest of all, arguing that religious reforms ought to be accompanied by political reforms granting more rights to the peasants. In an effort to calm things down, Luther issued a pamphlet entitled "A Sincere Admonition by Martin Luther to All Christians to Guard Against Insurrection and Rebellion" and preached a series of sermons in Wittenberg. His "Letter to the Saxon Princes" made it clear that he regarded Karlstadt and Müntzer as dangerous revolutionaries, even though they had been inspired by his religious reforms. When the uprising known as the Peasants' War broke out, Karlstadt renounced violence, but Müntzer led a peasant army into battle against mercenaries fighting on behalf of the nobility in May 1525. He was defeated, tortured, and beheaded.

The extent to which Luther's successful campaign depended on printing has been the subject of much debate in recent decades. Some historians have favored the view that printing was a major cause of the Reformation—a view sometimes characterized as "without printing, no Reformation." Others are more skeptical, arguing that the impact of printing must have been limited, given the low level of literacy in Germany at the time, and that preaching played the more significant role in spreading Luther's message. The fact that Luther succeeded where previous reformers had failed does suggest that the printing press made a difference, but its use must be considered in context. Pamphlets provided a medium for the wide distribution of Luther's views, but the embrace of the Reformation in much of Germany depended on a combination of factors, including the spreading of Luther's ideas in both spoken and printed form and the reluctance of the secular authorities to be pushed around by the church (something that was not the case in other parts of Europe, where the Reformation was effectively suppressed). Revolutions are ultimately caused by underlying grievances: oppression, political dissatisfaction, anger at corruption, and so forth. New forms of media do not trigger revolutions by themselves, but they can make it easier for would-be revolutionaries to coordinate their actions, synchronize opinion, and rally others to their cause. Luther's use of pamphlets represents a pioneering example.

# CHAPTER 4

# POETRY IN MOTION: SOCIAL MEDIA FOR SELF-EXPRESSION AND SELF-PROMOTION

*What Printing-presses yield we think good store,*
*But what is writ by hand we reverence more:*
*A Book that with this printing-blood is dyed*
*On shelves for dust and moth is set aside,*
*But if't be penned it wins a sacred grace*
*And with the ancient Fathers takes its place.*
—John Donne

## THE FACEBOOK OF THE TUDOR COURT

IN 1535 MARGARET Douglas, the niece of the English king Henry VIII and a lady-in-waiting to his queen Anne Boleyn, became secretly betrothed to Thomas Howard, another young member of the Tudor court. Romantic liaisons arose frequently in the heady atmosphere of the court, where aristocratic young men and women worked in close proximity (and beyond the gaze of their parents) as attendants, tutors, and governesses. But the secret marriage contract between Lady Margaret and Lord Thomas, aged twenty and twenty-four at the time, threatened to be rather more consequential. After Anne Boleyn suffered a miscarriage in January 1536 the king concluded that she was unlikely to produce a male heir and set in motion a scheme to replace her with his new mistress, Jane Seymour. The queen was accused of adultery with several men, including her own brother, and was put on trial in May. Anne and five of her supposed lovers were found guilty, and all of them were executed. Henry declared Elizabeth, his daughter by Anne, illegitimate, as he had previously done for Mary, the offspring of his first marriage. As a result, Lady Margaret unexpectedly found herself first in line to the English throne. So when Henry learned of her clandestine betrothal in July 1536 he was furious, considering it an attempt on the throne by Lord Thomas. The king had the couple arrested and locked up in the Tower of London.

Lady Margaret fell ill and was moved to Syon Monastery, where she remained under arrest. During their incarceration she and Lord Thomas did what Tudor aristocrats always did when imprisoned: they wrote poetry. Specifically, they wrote a series of love poems to each other. These poems survive because they were written down in an unusual book, now known as the Devonshire Manuscript, which Lady Margaret, Lord Thomas, and the other young courtiers in their circle passed around as a means of exchanging poems, notes, and coded messages with each other. A community of at least nineteen separate users of this surreptitiously shared book can be distinguished from their handwriting. The book's main contributors were Lady Margaret and her close friend Mary Shelton, along with Lord Thomas and his niece Mary Fitzroy, though many others whose names are not known also participated in the conversation that took place within its pages. The book seems to have circulated within this circle for several years, but was used most actively between 1534 and 1539.

Of the book's 194 entries, the vast majority are poems or verse fragments, including poems written by the book's users, excerpts of fourteenth-century verse by Geoffrey Chaucer, and contemporary works by several of the most prominent poets of the Tudor court, transcribed from circulating manuscripts. In particular, there are a large number of poems by Sir Thomas Wyatt, one of the leading poets of his day, along with works by Sir Edmund Knyvet, Richard Hatfield, Sir Anthony Lee, and Henry Howard, the Earl of Surrey. The interactions among the book's users in and around these poems take the form of comments written in the margins, wordplay of various kinds, the reworking of existing poems to fit the circumstances of the writer, and the composition of many entirely original works. This modest-looking book provides a glimpse into the social network of its users and encapsulates the range of uses to which poetry was put in the Tudor court. Today teenagers can commune in secret on social networks, on Internet forums, and using mobile phones. The Devonshire Manuscript provided a similarly secluded social space in which the young courtiers could exchange messages with each other, beyond the gaze of the outside world.

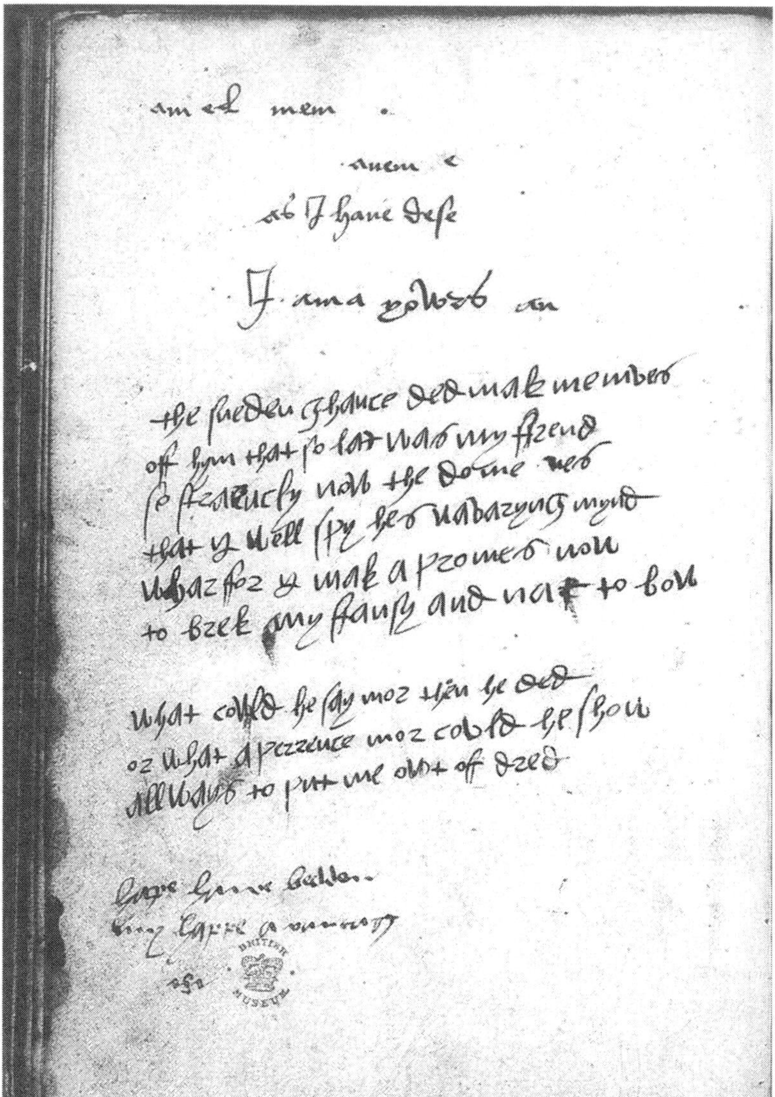

This page from the Devonshire Manuscript contains a cryptic anagram and a poem written in different hands, thought to be those of Thomas Howard and Margaret Douglas. The line "I am yowrs an" may refer to Anne Boleyn. *The British Library Board ADD MS 17492 f.67v/Adam Matthew Digital*

In one poem sometimes attributed to Wyatt, "Suffryng in sorow in hope to attayn," the writer declares his love for a woman whose name is artfully hidden in the first letter of each of the seven stanzas: SHELTUN. Evidently the poem was written by or on behalf of a suitor of Mary Shelton, who was romantically linked with many people, including Wyatt, Sir Thomas Clere, Henry Howard, and even Henry VIII himself. She shared the poem with her friends by copying it into the book, prompting one member of her circle (probably Lady Margaret) to write "fforget thys" in the margin next to it, to which Shelton responded "yt ys worthy." Even if she judged the poem itself to be worthy, however, she appears to have rejected the suitor. Below the poem she added the lines "ondesyard sarwes / reqwer no hyar / Mary Shelton" (undesired sorrows require no ear).

In several places Margaret Douglas and Mary Shelton appear to offer each other support by writing poems with similar themes next to each other, on subjects including professing devotion to a lover in the face of adversity, expressing the difficulty of having to conceal one's true feelings, and lamenting the happier times of days gone by. On one page an unknown hand adds an "amen" beneath a poem about lovers who are kept apart by others, which is perhaps a reference to the sorry tale of Lady Margaret and Lord Thomas; on another the comment "In the name of god amen" is written above Lady Margaret's transcription of a poem by Wyatt that deals with love, betrayal, and misfortune.

Mysterious signatures appear beneath two unattributed poems on facing pages, marked "first" and "second." In the first, "My ferefull hope from me ys fledd," the narrator explains that his "ferefull hope" that his affections will be returned by the woman he loves has given way to "faythefull trust" that she will accept him. In the second poem the suitor is advised not to put his trust in things he cannot see. The first poem is cryptically signed "fynys qd nobodyy" (the end, said nobody) and the second "fynys qd somebody" (the end, said somebody). Presumably the users of the book would have been able to relate these verses to an intrigue in the court or within their own circle; acting as a surreptitious channel for gossip and flirtation among its users was one of the book's functions. Another curiosity is a poem signed "qd 5813," which may be a cipher for "Wiat" (Wyatt). And one page contains a four-line riddle that is sometimes taken to be an anagram sent by Anne Boleyn to Wyatt, since it ends "I am yowrs an."

But as with so much else in the Devonshire Manuscript, it is hard to tell, nearly five hundred years later and divorced from the specific context in which it was written, what particular entries and notes mean.

By far the most substantial conversation recorded in the book's pages is the exchange of poems by Lady Margaret and Lord Thomas during their captivity. These poems seem to have been copied into the book by Mary Shelton, who may have acted as an intermediary between the lovers, since her father was a palace guard and her brother worked as a porter at the Tower. The couple's precarious situation before their arrest would have been known to the others in their circle, and the wider sharing of their poetic exchange through its inclusion in the book revealed their unhappiness and mutual devotion to their friends as well. In the early days of their captivity, at least, the lovers may even have relished the opportunity to assume in real life the classical roles of the star-crossed lovers depicted in courtly poetry, separated by the vicissitudes of fortune and bemoaning their plight in verse.

Lord Thomas begins the sequence with a poem lamenting "Alas that euer prison stronge / sholde such too louers seperate / yet thowgh ower bodys suffereth wronge / ower harts shalbe off one estate." In subsequent poems he expresses steadfast love for his "swete wyfe" and his optimism that his captivity will be brief and make him stronger. He compares himself to "a hawke that getes owt off hys mue" (a mew was a cage in which a hawk was kept while it shed its feathers). Lady Margaret responds that Lord Thomas is "the faythfullyst louer that ever was born," given the "great paynes he suffereth for my sake / contynnually both nyght and day." She insists her love for him "wyll not decay."

The mood in subsequent poems quickly darkens, however. Lord Thomas was told that he would only be freed if he renounced his love for Lady Margaret, something he refused to do—indicating that their betrothal was not, in fact, an attempt on the throne but a genuine attachment. He was condemned to death for committing treason, and a new law was hastily passed making it illegal for any member of the royal family to marry without the king's permission. In a gloomy poem Lord Thomas bids "adieu my none swete wyfe from T. h." The initials in the book were added later in another hand, indicating that a gap had been left when the poem was originally added to its pages. The sequence of poems ends with Lord Thomas quoting a passage from Chaucer's "Troilus and Criseyde," another tale of star-crossed lovers, in which a despondent Troilus expresses his desire to die so that his soul, at least, can be with his beloved. Again, in two places Lord Thomas misses out the name Crisyede and leaves a gap, and the poem's rhythm invites the reader to substitute the name "Margaret" in its place. The final lines of the sequence consist of another quote from "Troilus and Criseyde" in which Troilus asks those who are lucky in love to remember him when they pass his tomb: "But whan ye comen by my sepulture / remembre that yowr felowe resteth there / for I louyd eke thowgh I vnworthy were." Evidently Lord Thomas wished these words to be his epitaph. In the event he escaped execution, but the lovers were never reunited. Lord Thomas died of illness in the Tower on October 31, 1537, two days after Lady Margaret's own release.

## THE USES OF POETRY

The young users of the Devonshire Manuscript were by no means unusual in their love of poetry. It offered members of the Tudor court a parallel medium of communication in which the unsayable could be expressed in veiled or coded form. The circulation of poems in manuscript form provided a gossipy back channel behind the outward formality and strict rules of court life, and a conveniently ambiguous way to make political points. Poems were read out in small gatherings known as "pastimes," given as gifts, enclosed in letters, posted conspiratorially into people's pockets, or simply left lying around where they would be found and read. They would then be passed from hand to hand, copied and shared, and transcribed into notebooks (such as the Devonshire Manuscript) for future reference. Although individual poems might appear to describe a universal sentiment—the pain of love denied, the fickle nature of fortune—they were often written in response to specific goings-on at the time, and the joke was to tease out this hidden meaning. The more cleverly it was hidden, the better.

Paying compliments, making amorous advances, expressing grief or unhappiness, cracking jokes at the expense of others—all could be accomplished through poetry. Poems were a vital currency in the courtly "game of love," an elaborate parlor game that required ladies and their male admirers to conform to particular roles—she the unattainable and chaste object of desire, he the suitor who strives to prove himself and win her affection, even as he knows he must be refused. The fact that such playacting was sometimes used as camouflage for genuine affection merely added to the fun. Was a lover's lament merely adherence to a familiar poetic form, or an expression of real feeling? When a poem describing a particular situation was read out, who blushed? Who exchanged glances with whom? The ability to read and write poetry was a vital skill for members of the court, without which they risked being left hopelessly in the dark, unable to follow what others thought of the latest goings-on. Consider this poem, which circulated in the court in the early 1530s, during the agonizing period when Henry was trying to divorce Catherine of Aragon in order to make Anne Boleyn his queen.

*Grudge on who list, this is my lot:*
*Nothing to want if it were not.*
*My years be young even as ye see.*
*All things thereto do well agree.*
*In faith, in face, in each degree*
*No thing doth want, as seemeth me,*
*If it were not.*

The poem is the lament of a woman who is prevented from being with the man she loves by an impediment that is never specified. But the refrain "Grudge on who list, this is my lot" reveals its meaning: it is a translation of Anne Boleyn's French motto "Ainsi sera, groigne qui groigne," revealing the impediment to be Henry's marriage to Catherine of Aragon. The poet winks at the reader in the last few lines: "And here an End: it doth suffice/To speak few words among the wise"—implying that those in the know will understand the hidden meaning. Whether Anne Boleyn commissioned this poem or even wrote or performed it herself is unknown, but she must certainly have approved of it.

Similarly, one of Wyatt's poems, "The pillar perish'd is whereto I leant," which is outwardly an expression of anguish at the death of a lover, is widely taken to be a veiled denunciation of the execution of his friend Thomas Cromwell, one of Henry VIII's most trusted advisers, who fell from grace in 1540. Wyatt was an expert at exploiting the discretion and ambiguity afforded by poetry to express political dissent.

During his career as a diplomat in Henry VIII's court he also fell from favor on several occasions, and was imprisoned in the Tower for a while in 1536, possibly because of his rumored affair with Anne Boleyn, whose eventual execution he witnessed from his cell. But unlike many of his contemporaries, Wyatt survived the swift and dramatic reversals of fortune common to Tudor court life, dying of illness rather than at the executioner's block. This made him keenly aware of the courtier's predicament: the closer to the seat of power one gets, the more one must suppress one's own opinions, and the more dangerous it becomes to let the mask slip. It is telling that his satirical poem on court life, "Mine own John Poyntz since ye delight to know," was the most recopied item in surviving manuscripts from Henry's court, which indicates the extent to which others could relate to it. It takes the form of advice from a veteran courtier who has withdrawn from court life, exhausted by the need to pander sycophantically to those in power.

With convenient ambiguity, the poem is outwardly about the European courts that Wyatt had visited as an ambassador, but it can also be read as a description of Henry's court. The poem likens court life to being imprisoned in chains of gold ("Who so ioyes such kinde of life to holde / In prison ioyes, fettred with cheines of gold") and refers to the arbitrary exercise of power ("The powar of them, to whom fortune hathe lent / Charge over vs, of right, to strike the stroke"). The poem decries the immorality and corruption of the court, but it chiefly objects to the dishonesty of courtiers who go along with it to further their own careers. The humiliation suffered by courtiers obliged "on othres lust to hang boeth nyght and daye" would have been all too familiar to members of Henry's entourage.

Wyatt's poem is modeled on a satirical Italian poem of the same period and also draws upon the Third Satire by the Roman poet Juvenal, advising a friend what to do in Rome (and so giving the poet the opportunity to satirize the goings-on there). Translating existing poems provided further opportunities for sending subtle messages, given the conveniently large gray area between recycling specific themes (as Wyatt did with Juvenal) and presenting a faithful translation of an entire work. For a start, the choice of the text to be translated was itself meaningful; a poet could pick a work that matched the sentiment he wished to convey. The translation could then be subtly manipulated to make a particular point. And any such manipulation would be apparent only to readers familiar with the original text, thus hiding a further message for those in the know. If anyone took offense, the poet could always blame the author of the original text.

Such machinations are evident in Wyatt's reworkings of the Psalms, which appear on the surface to be religious poetry, but have a clear political subtext. The setting is the biblical court of King David, a tyrannical king led astray by lust. He is willing to kill a loyal subject, Uriah, in order to pursue his wife, Bathsheba. When another of his subjects reveals this state of affairs to all, the king is forced to acknowledge and apologize for his wrongdoing. These poems were written in the 1530s, probably during one of Wyatt's spells in the Tower. They were circulated, like all of Wyatt's poems, in manuscript form, only appearing in print after both Wyatt and Henry VIII had died. The contemporary parallel was rammed home by an introductory sonnet that Henry Howard, the Earl of Surrey, another poet-aristocrat, wrote to accompany Wyatt's work, which was circulated with it. Surrey expressed the hope that "rewlers may se, in a myrrour clere / The bitter frewte of false concupiscence" and that this "myght them awake out of their synfull slepe."

In a similar vein, Surrey himself produced a pioneering translation of books two and four of Virgil's *Aeneid* into English blank verse in the 1530s. This apparently odd selection (just two books from a work that was already available in English anyway) conveniently allowed him to explore the personal and political consequences of lust: the destruction of Troy by the Greeks in retaliation for the kidnapping of Helen, and Aeneas' abandonment of Dido, the queen of Carthage, prompting her suicide. But remember, folks, it's only a translation! The circulation and sharing of works like these within the court, and the selective quotation of relevant lines, gave courtiers a thrilling and illicit means of expressing their unease at the king's behavior and their recognition of their shared plight. On the surface, they were simply exchanging poems with their friends. But these person-to-person exchanges of poetry allowed subtle messages to be sent and received.

## MANUSCRIPT NETWORKS IN THE AGE OF PRINT

The circulation of poetry within court circles was just part of the much wider practice of exchanging and collecting poems and other manuscript texts with friends and family members that had arisen by the middle of the sixteenth century. The printing press was now a century old, but rather than making obsolete the copying and sharing of documents in manuscript form, print actually increased the importance and prevalence of handwritten text. Printing pushed up demand for paper throughout Europe, encouraging production and making it cheaper (its price fell by 40 percent during the fifteenth century) and more widely available. Printed books promoted literacy and writing manuals could be produced in quantity. The popularity in Tudor England of books explaining how to write letters illustrates that this was a new skill for many, which they were eager to learn. As more people learned to read and write, the circulation of handwritten documents of all kinds took off. Copying and recopying documents by hand, impractical in the era of parchment, became paradoxically widespread in the era of print. And once a document had been copied, it could be shared with or passed on to someone else. The resulting networks of manuscript circulation could then be used as means of self-improvement (by gathering educational or motivational texts), self-expression (what one passed on to others was a reflection of one's own character and interests), and self-promotion (as poems and other texts acted as a form of social currency between patrons and their protégés).

This activity centered around large notebooks, called miscellanies, anthologies, or commonplace books, into which noteworthy items were copied, in whole or in part, to create a personal trove of valuable information. Such books were mostly owned by individuals but were on occasion shared within families or groups of friends, as in the case of the Devonshire Manuscript. New items were entered in a miscellany in no particular order, in any available space, whereas commonplace books were more organized, with entries entered under specific headings to aid subsequent retrieval. In 1512 Erasmus of Rotterdam suggested detailed guidelines for keepers of commonplace books, suggesting that they should file noteworthy quotations under various headings for content, structure, or style. (Like other scholars at the time, he found the keeping of commonplace books to be a helpful antidote to the feeling of information overload as thousands of new pamphlets and books spewed from Europe's printing presses.) Once squirreled away, interesting facts or turns of phrase could be easily retrieved by the book's owner when composing his or her own poems, letters, or speeches. A commonplace book or miscellany also provided a record showing how its owner's interests or passions changed over time. The custom of keeping commonplace books in this way grew out of the earlier tradition among churchmen of collecting *florilegia* (literally, "gatherings of flowers"), which were compilations of excerpts from Christian and classical sources to be drawn upon when composing a sermon.

Miscellanies and commonplace books that survive from the Tudor period contain a huge variety of texts, including letters, poems, medical remedies, prose, jokes, ciphers, riddles, quotations, and drawings. Sonnets, ballads, and epigrams jostle with diary entries, recipes, lists of ships or of Cambridge colleges, and transcriptions of speeches. Collecting useful snippets of information so that they could be easily retrieved when needed, or reread to spark new ideas and connections, was just one of the many overlapping functions of commonplace books. The act of transcribing poems, quotations, or other information into a book would help the book's owner to remember the text in question. As Erasmus put it, "This method will also have the effect of imprinting what you read more deeply on your mind, as well as accustoming you to utilizing the riches of your reading."

At the same time, the practice of maintaining miscellanies and commonplace books and exchanging texts with others served as a form of self-definition: which poems or aphorisms people chose to copy into their books or to pass on to their correspondents said a lot about them, and the book as a whole was an expression of its owner's character and personality. People would sometimes lend their commonplace books or miscellanies to their friends, who could then page through the entries and copy anything of interest into their own books. Like Internet users setting up blogs or social-media profiles for the first time, compilers of such books seem to have relished the opportunities their newfound literacy gave them for projecting a particular image of themselves to their peers. A minority of the texts that people circulated were original compositions; most material was quoted from other sources. The same is true of modern social-media systems: posting links and snippets found elsewhere is standard practice on blogs, Facebook, and Twitter: and on some platforms, such as Pinterest and Tumblr, more than 80 percent of items shared are "repins" or "reblogs" of items previously posted by other users. Then as now, people enjoy being able to articulate their interests and define themselves by selectively compiling and resharing content created by

Eternall word, but these can not be prest
his greatnes wth such easy readines
and therefore yeeld, ye heaven shall pas away
ye sunne & moone & starres shall all obey
to light one generall bonfire, but his word
his Bible & his all destroying sword
that still survives, no wt of it can be
Each little minute immortality.
the worlds owne mother on whose breast did hang
the worlds upholder drawne into a span.
she, she was not so blest because she bare him,
because her selfe was new borne, & did heare him
before she had brought to, she heard her son
first speake in ye annunciation.
And yen, even then, before she brought to this
by name of blessed she her selfe inscribes.
Once more this mighty word his people greete
thus lapt, & thus swath'd up in paper sheete
read here gods image wth a zealous eye
the legible and written deity.

On one that died of an impostume in the head.

Is death so cunning now yt all her blows
aime at the heads? doth now her wary bow
make surer worke yn heretofore the steele
slew stout Achilles only in the heele.
Now finds out sleights, when men ythem selves begin
to bee their [illegible] false by new found sin.

Tis cowardice to make a wound so sure
Need art in killing, where no art can cure.
was it for hate of learning yt she smote
this upper shop where all ye Muses wrought
learning shall crosse her drift, & dayly buy
all wayes & meanes of immortality.
because her head was crusht doth she desire
Our equall shame, in vaine she doth aspire
no, no we know where ere she make a breach
her poysnous sting only ye heele can reach.
look on ye soule of man, ye very heart
the head it selfe is but a lower part.
yet hath she strayned her utmost tyranny
and don her worst in that she came so high.
had she reservd this stroke for haughty men
for politique contrivers, justly then
the punishment were matcht wth ye offence,
but when humility & innocence
be indiscreetly in the head are hit
death hath don murther, & shall die for it.
thinke it no fault shewne, because ye braine
the seat of sence, & yen more free from paine
thinke it no kindnes, wn so stealingly
he rather seemd to shrinke away yn die
and like ye innocent, ye widowes children
cride out, my head, my head, and died.
thinke it was rather double cruelty
slaughter imbrued on his name, yt he
whose thoughts wer nothing tainted, nothing against
might seeme to hide corruption in his braine.

A page from a shared seventeenth-century commonplace book, containing poems by multiple authors in various hands. *James Marshall and Marie-Louise Osborn Collection, Beinecke Rare Book and Manuscript Library, Yale University* others. The mere act of sharing something can, in other words, be a form of self-expression.

Commonplace books and miscellanies preserve traces of the sharing and copying system that linked their users together as they traded texts. Poems were circulated on loose sheets, some of which survive in miscellanies, though it was common practice to retranscribe an incoming poem, rather than binding or sticking the sheet on which it was received into one's book, so that the original could then be returned to its owner or forwarded to someone else. Longer works of prose or collections of poems might be passed around in the form of a quire—several sheets of paper folded and sewn together to make a booklet. Even quite long prose works circulated in manuscript in this form, such as Sir Philip Sidney's *Arcadia* and Sir Walter Raleigh's "A Dialogue Between a Counsellor of State and a Justice of the Peace," indicating a willingness among readers to copy out lengthy documents.

In some cases marginal notes indicate where a particular item came from, explain how it reached the transcriber, or offer instructions to the borrower, such as "Mr Thomas Scott sent these verses by the hand of Dr John White to Sr Walter Raleigh," "A farewell to desire geven by J.T.," "Brother Vessy Matt Bacon desires you to send this up againe to his mothers, when you have sithen usd it, or writt it out. JT," "the manuscript of Judge dodridge will shew you more." But usually the source is unmentioned and may even be obscured. On the back of an obscene poem in one manuscript collection now in the British Library is written "Pray my Lord tell nobody from whom this Song comes, for I am ashamed to owne it." The similarities and overlaps between several manuscript collections compiled at the University of Oxford indicate widespread sharing of both individual texts and entire miscellanies among students and their tutors. The universities, the Inns of Court in London (the center of the legal profession), and the royal court were the most active centers of manuscript exchange, with texts flitting between students, lawyers, and courtiers, respectively. Manuscripts also traveled along social networks associated with families, political factions, or religious groups.

For ambitious young men, participation in these manuscript-exchange networks provided an opportunity for self-promotion and social advancement, by giving them an excuse to maintain regular contact with their patrons. Poems and other texts served as a form of social currency that could be used to establish and maintain useful connections. Those skilled at writing their own poetry could dedicate works to influential patrons in the hope of winning their favor and having their works passed on to other influential people. The efficiency of the manuscript-circulation system meant that English poets addressing particular audiences of patrons or fellow courtiers could rely entirely on manuscript transmission; there was no need to resort to print. Many of the most influential poets of the period, including Thomas Wyatt, Philip Sidney, Walter Raleigh, and John Donne, circulated their work almost exclusively in manuscript form. This is sometimes taken to indicate an elitist disdain for print as an unworthy, grubbily commercial medium. But the simple fact is that their poems were primarily intended for specialized groups of readers who could be reached easily through scribal transmission alone; it was a question of horses for courses.

Admittedly, poets' works sometimes broke out beyond the small coteries for which they were originally written, and even into print. Philip Sidney never had any of his writings printed and told his sister that he had written *Arcadia*, which circulated widely in manuscript, "only for you, only to you." But by the time of his death in 1586 his lyric poetry was also readily available across the Tudor document network; his works were more widely shared in manuscript form than those of any other English poet. Similarly, William Shakespeare distributed "sugard sonnets among his private friends" during the 1590s, but they quickly leaked out of his immediate circle. His sonnets were referred to in a book published in 1598, and two of them were included in a printed anthology, "The Passionate Pilgrim," in 1599. John Donne referred to copies of some of his poems "which have crept out into the world without my knowledge." He and most of his contemporaries were not interested in addressing a wider public. They regarded poetry as a way to amuse friends, win the favor of patrons, and advance their careers.

## THE TWEETS OF @SAUCYGODSON

Poetry continued to be important in the court of Henry VIII's daughter, Elizabeth I. Like her father she enjoyed witty poetic banter and wrote poetry herself (composing some of it while being held prisoner in her youth). In Elizabeth's court, a talent for poetry offered an aspiring young courtier the opportunity to make his name and win advancement. Today some people manage to exploit their prowess at blogging or tweeting to achieve fame, win a plum job, or land a book deal. Similarly, the ability to write poems, coin epigrams, and respond to others with witty put-downs provided a means for young men in the Elizabethan era to demonstrate the breadth of their knowledge and the quickness of their wit, and by extension their suitability for appointment to official posts—as the career of Sir John Harington illustrates.

Harington's father, also called John, had served in Henry VIII's court as an attendant to Princess Elizabeth, as she then was, and the younger John ended up as one of the childless queen's 102 godchildren. A brilliant student, he inherited a love of poetry from his father (a collector of many poems by Wyatt and Surrey) and demonstrated his ability to write epigrams in Latin while still a schoolboy at Eton. After completing a master's degree at Cambridge University and studying law in London, he appeared at the court for the first time in 1581 at the age of twenty-one, and his witty epigrams, this time in English, immediately brought him to prominence. They were satirical and daring in their humor, but their chief purpose was to advertise the dazzling intellect of their author. Harington had a witty riposte to hand in any situation, often with a provocative hidden meaning. Elizabeth encouraged him, her indulgence of his risqué humor indicating to her court that she was a more tolerant ruler than her father had been. A portrait from the early 1590s shows Harington with a white ruff collar, an elegantly pointed beard and moustache, and a mischievous smile playing on his lips. Harington soon became known as the queen's "saucy godson" for quips like this one:

*Treason doth never prosper: what's the reason?*
*Why, if it prosper, none dare call it treason.*

This was vintage Harington: a good joke masking a deadly serious point. His father had endured a few spells in the Tower and Harington was aware just how dangerous the life of the courtier could be. His response was to play the part of the wise fool, jesting on the sidelines of Elizabeth's court and wrapping up his moral and political barbs in apparently harmless witticisms, their true meaning only apparent once the laughter had subsided. He followed the advice of Castiglione, author of *The Book of the Courtier*, a popular guide to court etiquette at the time, who observed that "to be in a mask brings with it a certain liberty and licence," enabling the witty courtier to surprise his listeners "when they behold afterward a far greater matter to come of it than they looked for under that attire." Harington exploited his position as one of the queen's favorites to make veiled attacks on the corruption of her officials and clergymen. He even went so far as to criticize her father for his unfortunate habit of having his wives beheaded. In one of his epigrams, a noblewoman receives an invitation to marry the king, but declines:

*. . . I greatly thank the king your master,*
*And would (such love in me his fame hath bred)*
*My body venture so: but not my head.*

Such quips were eagerly whispered from one courtier to another and circulated in written form within the court and beyond. Harington himself gave manuscripts of collections of his epigrams to close friends and family members. Once poetry had been employed to win a place at court, it could then be used to ask for advancement or, in the event of falling from favor, to apologize for misdeeds. One of Harington's poems, which he wrote for the queen and left "behinde her cushion at my departing from her presence," explicitly links poetry with political advancement. He flatters the queen that her careful reading of one of his poems greatly improved it, and asks her to do the same for his personal circumstances, reading them closely and in the process improving them: "Like as you read my verse, so—read my Fortune," he jokingly asks, signing the poem "From your Highnesse saucy Godson."

Harington was duly given some minor official duties, but after his promising start he overstepped the mark when he translated part of "Orlando Furioso," an Italian epic poem, into English and circulated it at court. The translation was of the poem's raciest section, a saucy story-within-a-story about love, sex, infidelity, and a queen who has sex with a dwarf. When the queen discovered her ladies in waiting passing around a manuscript copy of this indelicate tale, Harington found himself in hot water. But rather than banishing him from the court completely, Elizabeth devised a suitable punishment for him.

She said he was not to return until he had translated the vast poem in its entirety—something she assumed he would be unable to do. Harington proved her wrong, eventually returning to court with a full translation of the poem that he dedicated to the queen, regaining her favor in the process.

It was not long before he was in trouble again, however. In 1596 he published a curious scatological work entitled "The Metamorphosis of Ajax," which uses the unlikely topic of household sanitation as the basis for an elaborate political allegory. The title is a pun on "jakes," Elizabethan slang for toilet, and the "metamorphosis of a jakes" that Harington describes is his new invention: the flushing toilet. (His name is sometimes said to explain the use of the word *john* to mean lavatory.) Having described its operation, Harington considers "how unsavoury places may be made sweet" in a broader sense. He draws an analogy between a clean home, kept free of disease with the help of his invention, and the need for similar cleanliness in politics. He decries the "stercus" or excrement poisoning society, in particular the political attacks being made on his uncle at the time. These political references landed Harington in trouble with the queen once more, and he was banished from the court again "til he had grown sober"—though the queen did agree to have one of Harington's "water closets" installed at her palace in Richmond. Harington later ended up with a knighthood, conferred upon him by the Earl of Essex during a disastrous military campaign in Ireland—which invoked the queen's anger yet again. But as ever Harington's wit saved him, and he remained a favorite of the queen's until her death in 1603.

Harington's use of daringly witty poetry to draw attention to himself inspired other ambitious young men, who lacked his family connections, to do the same. Sir John Davies, who was nine years younger than Harington, made his name with some even sharper epigrams, winning the support of the queen and embarking on a successful legal and political career. (Some of his poems contain the name "Elisabetha Regina" hidden as an acrostic.) When Elizabeth died, Davies was a member of the group sent to bring King James VI of Scotland to London to become the new king; and during his reign (as James I of England and Ireland) Davies used his poetry to win further political preferment. John Donne also followed in Harington's footsteps; his poetry, starting with epigrams, won him an appointment as secretary to the Lord Keeper of the Great Seal, Sir Thomas Egerton. But Donne then wrecked his political career by marrying Egerton's niece without permission. He continued to use his poetry, passed around in court circles, in an effort to attract political patronage, but without success. Eventually he was ordained as a priest and became well known as a preacher as well as for his poetry.

Yet neither Harington, Davies, nor Donne had set out to be remembered as a poet; for all of them poetry was a means to an end. Donne wrote to a friend in 1609 that he sought a more serious career ("a graver course"), quoting a Spanish proverb that "he is a fool which cannot make one Sonnet, and he is mad which makes two." But putting clever poems into circulation, like the canny use of social media today, could launch a young man's career and act as a stepping-stone to other opportunities. The use of social media for self-expression and self-promotion is nothing new, but dates back at least as far as the Tudor court of the sixteenth century.

# CHAPTER 5

# LET TRUTH AND FALSEHOOD GRAPPLE: THE CHALLENGES OF REGULATING SOCIAL MEDIA

*Give me the liberty to know, to utter,*
*and to argue freely according to conscience, above all liberties.*
—John Milton, *Areopagitica*

## A MESSAGE FROM THE HAND OF JOHN STUBBS

ON NOVEMBER 3, 1579, John Stubbs walked up onto the scaffold in the marketplace in Westminster to receive punishment at the executioner's block. Stubbs was an English lawyer of Puritan bent who had written a pamphlet entitled "The Discoverie of a Gaping Gulf Whereinto England is Like to Be Swallowed by Another French Marriage." It argued forcefully against the proposed marriage between Queen Elizabeth and the brother of the King of France, the Duke of Alençon, who was accused of being the devil "in shape of a man." Whether Elizabeth ever truly considered marrying the duke, to whom she gave the affectionate nickname "frog," is unclear, but she certainly wished to establish closer links between England and France as a counterweight to the might of Spain. The prospect that the queen might marry a Catholic alarmed English Protestants, such as Stubbs, who worried that she might also be planning to restore Catholicism as England's state religion. Stubbs considered himself a patriot defending his country, and his pamphlet attacking the duke, printed in secret and circulated throughout the country, reflected the views of many of his countrymen, not to mention much of Elizabeth's own court, which eventually persuaded her to abandon the idea of the marriage.

The queen was furious at Stubbs's impertinent meddling in both her private life and the delicate negotiations with the French, and she responded to his pamphlet by issuing a printed proclamation condemning his "lewd, seditious book, of late rashly compiled and secretly printed and afterwards seditiously dispersed into sundry corners of the realm." The pamphlet was denounced for containing "false statements" and "manifest lies," and although wiser readers would, of course, perceive this at once, there was a danger that "the simpler sort and multitude" might be led astray by it. There was no question of any change to English law or religion as a consequence of the proposed marriage, the proclamation went on, concluding that "Her Majesty willeth and straitly chargeth that both the foresaid book or libel, wheresoever they, or any the like, may be found, shall be destroyed in open sight of some public officer." Within days of the proclamation's appearance Stubbs had been arrested, tried, and found guilty of breaking a law "against the authors and sowers of seditious writings." He was sentenced to have his right hand, with which he had written the offending pamphlet, cut off with a cleaver.

Standing on the scaffold, Stubbs gave a short speech to the somber crowd that had gathered, insisting that he had not meant to offend the queen, and grumbling that she had pardoned others who had committed far greater crimes, yet had refused his plea for clemency. He then knelt down, placed his hand on the block, and uttered a dreadful pun: "Pray for me, now my calamity is at hand." With three blows of a mallet on the cleaver his hand was severed. To emphasize his loyalty to the queen, Stubbs removed his hat with his left hand and managed to cry "God save the queen" before fainting. The wound was bandaged and Stubbs lived for another twelve years, eventually dying of old age after a long and varied career as a lawyer and public servant.

The tale of John Stubbs's hand illustrates that European rulers had realized, in the wake of Martin Luther's pamphlet campaign and the Catholic church's ineffective initial response to it, that attacks on authority that appeared in print and attracted widespread attention had to be swiftly answered in kind. But ideally, they felt, such responses should not be necessary at all. They set about imposing strict controls on printing in the hope of preventing dangerous religious or political texts from appearing in the first place. And if someone got around the system, as Stubbs did by having his pamphlet printed in secret, then meting out the occasional brutal punishment would send a clear message to anyone else thinking of doing the same. Imposing regulations on printing was, in a sense, just bringing it in line with other professions, such as baking and weaving, which were already subject to strict rules. As Erasmus had noted in 1525, "not everyone is allowed to be a baker, yet making money by printing is forbidden to nobody."

Accordingly, over the course of the sixteenth century increasingly strict controls were imposed across Europe on what could be printed, and by whom. The preferred approach was prepublication licensing, which required the publisher of a text to submit it to the religious or secular authorities, or both, for approval. Some texts would be approved, others rejected outright, and in some cases approval would be granted on the condition that certain changes were made. In theory, only approved texts could then be printed. At the same time, there were attempts to outlaw clandestine printing, ban prohibited books, and restrict imports of books printed elsewhere. Printers were forced to join guilds so that they could be more easily monitored, and printing presses operated by nonmembers were seized and destroyed. French rules specified that "printing is not to be carried out in secret and hidden places." Venetian printers had to license each text with religious magistrates and the printers' guild. All printing presses in Saxony were ordered to be shut down except for those in Wittenberg, Leipzig, and Dresden, and printers had to register and swear an oath of loyalty to the duke. In England licensing was made the responsibility of the bishops, and the Worshipful Company of Stationers, which had been granted a monopoly on printing, was given increasingly draconian powers to search for seditious books and destroy unregistered presses. The company argued that "if every man maie print, that is so disposed, it may be a meanes, that heresies, treasons, and seditious Libelles shall be too often dispersed, whereas if onlie knowne men doo prynte this inconvenience is avoided." Its members supported a strictly controlled media environment because it protected them from competition from unlicensed printers, allowing them to keep prices high. But the company defended this self-serving position as being in the best interests of society as a whole.

Enterprising printers soon found ways around these rules, however. Venetian printers evaded local regulations by adding the words *con licentia* even to unlicensed works, or by labeling them as having been printed elsewhere, which meant they were not subject to local rules.

Printers elsewhere used the same tricks, giving false names and addresses in the imprints of controversial works and labeling them as licensed, even when they were not. Clandestine presses continued to operate, in some cases shifting their location from time to time to avoid detection. The authorities responded by passing even stricter laws. The Duke of Saxony decreed in 1571 that anyone printing or circulating unapproved works would be "relentlessly punished." In 1581, after an outburst of unlicensed Catholic pamphlets, the English Parliament instituted the death penalty for anyone who wrote, printed, or disseminated any work "containing any false, seditious, and slanderous matter to the defamation of the Queenes Majesty, that now is, or to the incouraging, stirring or moving of any insurrection or rebellion." A royal proclamation of 1588 required England's state officials to "inquire and search for all such bulls, transcripts, libels, books and pamphlets, and for all such persons whatsoever as shall bring in, publish, disperse, or utter any of the same." Printers and booksellers were subject to raids, arrest, fines, and harassment if they were suspected of breaking the rules. And the haphazard and capricious way in which the porous but occasionally brutal laws were enforced actually made them more effective. Never being quite sure whether one would get into trouble or not acted as a further deterrent against wrongdoing and encouraged most printers to play it safe.

The steady tightening of the rules can be seen most clearly in England, by considering the proportion of known printed texts (excluding official texts such as government proclamations) that were officially licensed before publication. The figure went from 3 percent in the 1560s, after licensing was first introduced, to 7 percent in the 1570s and 42 percent in the 1580s. It shot up to 87 percent in 1589–90 after a clampdown triggered by an outburst of Puritan pamphlets attacking the authority of the bishops, which appeared under the pen name of Martin Marprelate and were printed using a itinerant press moved around the English midlands. (The wife of one English noble even allowed the press to be set up in her parlor, assuring her husband that it was equipment for embroidery.) In the first three decades of the seventeenth century, under Elizabeth's successors James I and Charles I, around two thirds of printed texts were licensed. The figure rose to more than 75 percent in the 1630s, when censorship was tightened again under William Laud, the archbishop of Canterbury who, as England's most senior bishop, was required to oversee licensing. He took a particularly tough line against Puritan writers who dared to argue in print that the Church of England needed to be purged of the remaining vestiges of Catholicism.

Laud's most famous victim was William Prynne, a Puritan pamphleteer who had the tips of his ears cut off in 1634 for writing a work that was taken to be an attack on Henrietta Maria of France, the Catholic wife of King Charles I. Prynne was convicted for a second time in 1637, along with two other Puritan writers, John Bastwick and Henry Burton, for writing unlicensed books attacking the church. (Bastwick's book attacking the bishops was printed in the Netherlands and smuggled into England, and Burton had published an unlicensed pamphlet of sermons directly attacking Laud.) Bastwick and Burton had their ears cut off in the pillory, and Prynne had the remnants of his ears removed. For good measure he was also, at Laud's instigation, branded on his cheeks using "an iron made exceeding hot" with the letters "SL" for "seditious libel," the crime for which he had been convicted. That same year an edict was passed by the Star Chamber, the king's secretive and unaccountable high court, that printed books should not contain anything "that is contrary to Christian Faith, and the Doctrine and Discipline of the Church of England, nor against the State or Government, nor contrary to good life, or good manners, or otherwise, as the nature and subject of the work shall require." This gave Laud extraordinary powers of censorship. Yet it proved to be a high-water mark. A decentralized media system is very difficult to police effectively. It was not long before the licensing system broke down, and England embarked upon an unexpected and historic experiment in press freedom—the consequences of which reverberate to this day.

## AN APPETITE FOR NEWS

England's tightening controls over printed matter were prompted in part by growing political tension between the crown and Parliament. As in other European countries, initial concern that the press would be used to stir up religious unrest gave way to broader worries over its political uses. With print out of bounds because of the licensing rules, English political news began to circulate in large quantities via the existing manuscript-exchange networks instead, feeding a growing appetite for unofficial sources of news.

Officially it was illegal to report the proceedings of Parliament, in any form, outside its walls. Not even members of Parliament were meant to discuss politics except within Parliament itself and its various councils. No public discussion of politics was tolerated at all, and the idea that ordinary people could be trusted with political information was thought ludicrous. In practice, however, it was common for politicians to circulate manuscript copies of their speeches among their friends and associates, and for summaries of parliamentary proceedings to be jotted down and circulated. Such parliamentary reports, which had been doing the rounds since Elizabethan times, became far more numerous in the early seventeenth century as political tensions increased. The crown had the authority to intercept and censor letters but rarely did so because it suited the political elite, including members of the Privy Council that advised the king, to be able to keep up with political goings-on. So a blind eye was turned. Manuscript documents, unlike printed documents, were not considered to be public. They were assumed to be limited to the elite, who could be trusted with political information.

As with poetry, manuscript copies of speeches and parliamentary reports circulated as "separates" or "pocket manuscripts," which were self-contained texts passed and copied from one person to another. Exchange of manuscripts happened along existing social networks within court circles, the clergy, the academic community, religious sects, and political factions. On occasion political separates were produced in quantity: as many as fifty manuscript copies survive today of some items, which suggests that clerks or scriveners were employed to prepare lots of copies quickly. Many of the speeches that survive from the early seventeenth century are so polished that they cannot possibly be from transcriptions. They were probably put into circulation by the speakers themselves and may not resemble the actual speech delivered (if indeed it ever was) at all. Circulating a speech in manuscript form was a means of self-promotion and political activism, just as it had been in Roman times.

By 1600 an additional way of keeping up with the political scene had emerged in the form of regular bulletins, called "letters of news" or "news letters," compiled by political informants in London and sent to recipients in the country. A pioneer of this genre was John Chamberlain, a well-connected gentleman of private means who gathered news and gossip in London for his friends and then wrote them detailed letters explaining the latest goings-on. Chamberlain's letters, many of which survive, are regarded as an incisive and objective portrayal of London during the reign of James I. Having inherited a fortune from his father, Chamberlain did not need to work and wrote for his own amusement, without asking for payment. Not everyone was lucky enough to have such a willing informant, however, and other letter-writers such as John Pory and Edmund Rossingham subsequently began offering a weekly service to paying customers, at a cost of £5 to £20 a year. This was an impressive sum at the time (equivalent to £600 to £2,400, or about $1,000 to $4,000 today). With a few dozen subscribers, newswriters could make a good living.

At the time the prime location for gathering information was the old cathedral of Saint Paul's, where people met each day to exchange news and gossip. According to an account of such "Paul's walking" published in 1629, "The Noyse in it, is like that of Bees, a strange humming or Buzz; mixst of walking Tonges and feet; it is a kind of still roare, or Loud whisper. It is the great Exchange of all discourse, and no businesse whatsoever but is here stirring and afoot." This might seem an odd use of a cathedral, but Saint Paul's had fallen into disrepair, having lost its spire in a fire in 1561, and had come to be treated as a market and meeting place as well as a church. Repeated attempts by the church authorities to put a stop to such nonreligious uses failed. Saint Paul's was also the center of the London book trade, with many bookstalls in the surrounding yard and bookshops nearby offering newswriters further sources of information in the form of news ballads, books, and pamphlets, and gossip with shopkeepers and their customers. More gossip could be picked up around Parliament, at the Royal Exchange, the city's commercial center; from Thames boatmen (the equivalent of modern taxi drivers); in taverns; and from friends in high places. Newswriters would also try to get hold of copies of speeches, in some cases directly from their authors, and accounts of parliamentary proceedings, either derived from notes taken in the chamber or assembled from verbal reports.

All of this would be compiled into letters, in which tidbits of information were frequently preceded by the words "it is said," "they say," "I hear," "I have heard," and so on, emphasizing the oral nature of much of the news they contained. Newswriters usually wrote slightly different letters to each of their subscribers, tailoring the contents according to their needs and interests. Sometimes they did not write the entire letter themselves, but had scriveners copy out the bulk of the text, which was common to each letter, and then added comments and embellishments in their own hands. According to one account from the period, five scriveners employed by one newswriter each had to copy out sixteen letters every Tuesday, thirteen every Thursday, and fifteen every Saturday to meet the varied demands of more than one hundred subscribers. Even if a scrivener had copied out most of a news letter, the polite fiction was maintained that it was a private correspondence between two gentlemen, rather than a commercial transaction.

The expense of subscribing to a news letter probably explains why no effort was made to control them. They were assumed to have a small and limited circulation among the well-to-do. But they actually circulated more widely than this small group. Some recipients shared letters with their neighbors and split the cost of the subscription. Manuscript letters of news could be bought off the shelf from booksellers in London. News letters were frequently "chain copied" with either the original or a copy sent on to friends. Viscount Scudamore, a member of Parliament during the 1620s, collected copies of news letters he was sent by friends; he seems to have sent them copies of John Pory's news letter, to which he subscribed, in return. Sometimes newswriters exploited social ties between recipients as a means of delivery and distribution. Sir Simonds D'Ewes, an English politician, wrote a letter of news from London for his friend Sir William Spring, but rather than sending it directly he sent it first to Joseph Mead in Cambridge, who read it and passed it on to his friend and patron Sir Martin Stuteville (in Dalham, Suffolk, to the east of Cambridge), who in turn passed it on to his friend Spring (in Pakenham, a bit farther east still). For his part, Mead also subscribed to Pory's news letter, and wrote a regular digest of it and other sources for Stuteville, supplying his older patron with news updates, much as Caelius had for Cicero. And all these writers would have shared the letters they received, or relevant excerpts of them, with other correspondents as well.

The licensing system meant that some forms of news were allowed to appear in print, as long as they steered clear of the touchy subjects of domestic politics and religion. Accounts of miracles, disasters, murders, and monstrous births were particularly popular. In England, such accounts appeared first as single-sheet news ballads in the 1560s, though the ballad format gave way to the multipage pamphlet in the 1580s. Often illustrated with graphic woodcuts, these publications provided surreptitious opportunities to express concerns about moral decline or social unrest, or to imply that these were outward signs of deeper political or religious problems. The printing of foreign news was also permitted. Demand for foreign news surged after the outbreak in 1618 of what is now called the Thirty Years' War, which began as a religious war between Protestants and Catholics and eventually engulfed most of Europe. English Protestants wanted to know how their fellow Protestants were doing, fearing that a Catholic victory on the continent would be followed by an invasion of England.

Their desire for news about the war led to the appearance of a new type of publication: the coranto. This was a single sheet, printed on both sides, with a compilation of items, usually letters or eyewitness accounts of battles or other notable events, each of which was preceded by a heading giving the place and date of origin, such as "Out of Prague, the 5th of November." The first corantos (also known as "courants" or "corrants") appeared in the Netherlands in 1618, and they were soon being reprinted in English and sent to London, but with items about England omitted to avoid offending the authorities. This inspired Nathaniel Butter, a London stationer, to launch his own coranto, presented in pamphlet format rather than a single sheet, and based on material translated from German and Dutch publications. Printed corantos were much cheaper than manuscript news letters, costing two pence an issue (a weekly manuscript news letter cost its subscribers six to twenty-four times as much). With a typical print run of a few hundred copies, their circulation was larger, though not much larger, than that of manuscript newsletters.

Being anonymous, corantos were regarded as less trustworthy than handwritten news letters, which often related news heard at first hand. In 1621 the news-letter writer John Chamberlain complained about "corantos with all manner of newes, and as strange stuffe as any we have from Amsterdam." But rather than competing, the two formats proved to be complementary. Corantos could be enclosed within manuscript news letters as they circulated, providing printed foreign news alongside the handwritten domestic sort. Letters from this period contain abundant references to enclosed printed material. Mead's letters to Stuteville contain extracts from corantos, entire transcribed copies of them and, on several occasions, the printed corantos themselves. Corantos were printed compilations of what were originally manuscript documents, and the information they contained was in turn recycled into manuscript news networks. Print and manuscript reproduction worked together to accelerate the sharing of information.

This proliferation of new and untrustworthy news sources was not universally welcomed. The newswriters, coranto-makers, and news junkies they served were satirized by Ben Jonson in his play "The Staple of News," first performed in 1625, which contains many puns on the name "Butter." He and other newsmongers, Jonson implied, made up whatever news they thought their credulous customers wanted to hear. Like many of his contemporaries, Jonson thought it dangerous for ordinary people to have greater access to news, because printing allowed rumors and falsehoods to spread quickly, causing social and political instability. Moreover, the very existence of corantos had political connotations in England in the early 1630s, as Charles I did his best to stay out of the Thirty Years' War. He had fallen out with Parliament, and without its tax-raising powers he could not afford to participate in the conflict and instead made peace with France and Spain. The great interest in England in the plight of Protestant forces on the Continent, manifested in the popularity of corantos, was an implicit rebuke both of Charles's failure to intervene on their behalf and of his coziness with Catholic Spain. When Catholic forces suffered reversals that were reported in English corantos in October 1632, the Spanish ambassador in London complained. Charles duly asked the Star Chamber to ban all corantos, declaring that even foreign news was "unfit for popular view and discourse." Butter did not expect the ban to be permanent. He told his friend Pory that "he hoped ere long his Currantos would be revived." In the event, the ban was maintained for several years, and manuscript exchange once again became the dominant means of national news circulation. But Pory, too, felt that the ban on printed news publications could not last, writing to a friend that "they will burst out again one of these days." They were about to do so in spectacular fashion.

## THE EXPLOSION OF 1641

Charles I ruled without Parliament from 1629 to 1640, a period known as the "Personal Rule" by his supporters and as the "Eleven Years' Tyranny" by his enemies. He devised various clever schemes to raise money without recourse to Parliament, such as reinstating a long-forgotten law that wealthy landowners were required to attend the king's coronation, and then fining all those who had failed to show up, four years after the event. This and other tax measures were unpopular, but Charles could manage without Parliament provided he did not try to wage any expensive wars. In 1637, however, he and the hated Archbishop Laud made the mistake of trying to impose religious uniformity across England and Scotland, which triggered a Scottish rebellion. Charles reluctantly recalled Parliament in 1640, but it refused to provide the money for a military campaign, even though Scottish forces had occupied the north of England. The king's weakness was demonstrated when Parliament impeached and imprisoned his two closest advisers, the Earl of Strafford and Archbishop Laud. An act abolishing the court of the Star Chamber was passed, with the king's reluctant approval, in July 1641, greatly reducing the king's personal power. With Laud in the Tower of London and the Star Chamber abolished, the mechanisms for regulating the press had abruptly vanished, just as the growing political crisis had created immense demand for news. The result was an explosion.

Political and satirical pamphlets began to pour from London's presses. For the first time, printers dared to publish political speeches that had previously circulated only in manuscript form. By mid-November concern was being expressed in Parliament about the need to find "some meanes to restraine this licentious printing." In abolishing the Star Chamber, Parliament's aim had been merely to restrict the king's power, but it had the unintended side effect of removing the controls on the press. As Parliament's disagreement with the king intensified, its members concluded that it was in their interests for the public to learn more about their grievances. These were summarized in the Grand Remonstrance, a list of 204 complaints about the king's rule, which Parliament agreed to release in manuscript form and then in print, too. At the same time, detailed handwritten accounts of the proceedings in Parliament, under the title "Diurnall Occurrences or Heads of the Proceedings in Parliament," which had been compiled and circulated on a weekly basis since late 1640, also started to appear in print with the tacit approval of John Pym, Parliament's unofficial leader. These weekly printed accounts of political events, published every Monday with a summary of the previous week's proceedings, first appeared in late November. Known as newsbooks, they were the first publications to provide domestic political news in print. After decades of effort to prevent it, politics was suddenly being conducted and discussed in the open.

Nowhere was this more apparent than in the exchange of letters, in the form of pamphlets, between king and Parliament in early 1642 in which both sides appealed to public opinion, thus recognizing its importance and increasing its power. The printing of the Grand Re-

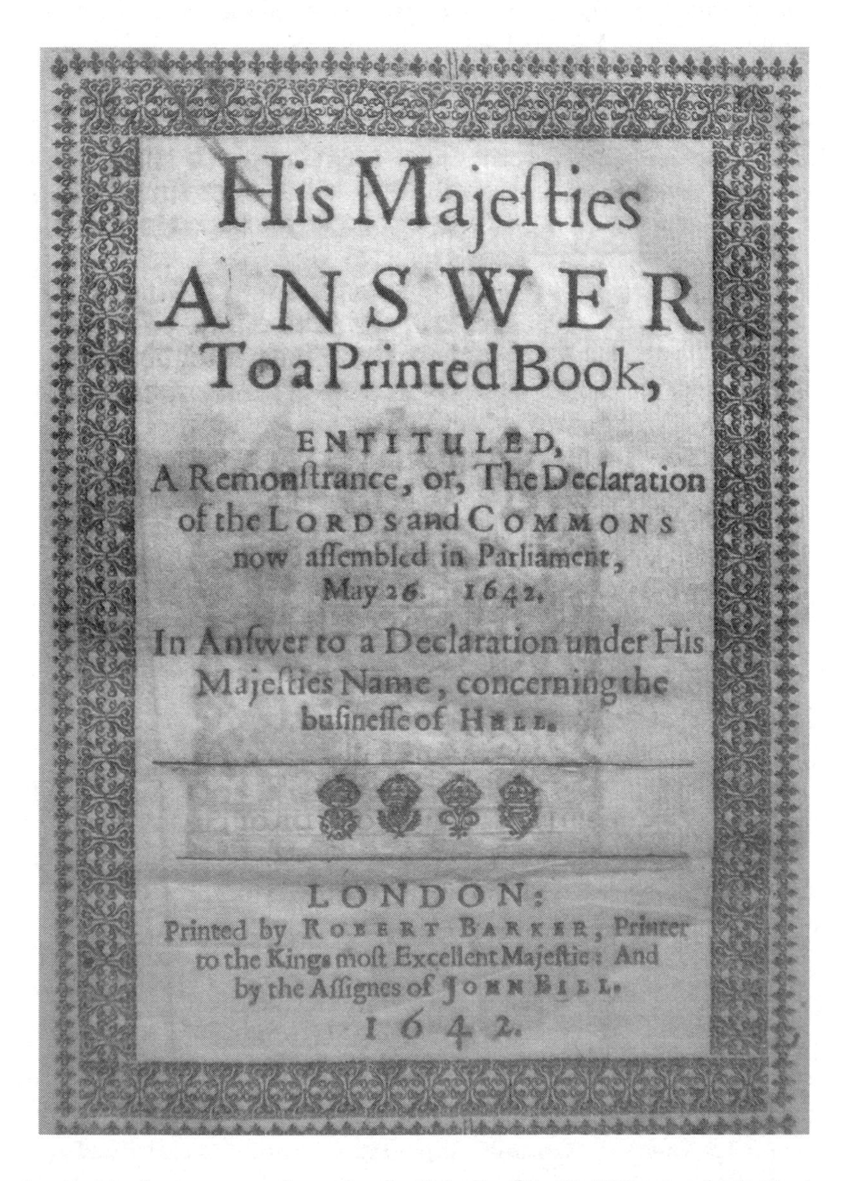

His Majeſties
ANSWER
To a Printed Book,
ENTITULED,
A Remonſtrance, or, The Declaration of the LORDS and COMMONS now aſſembled in Parliament, May 26. 1642.
In Anſwer to a Declaration under His Majeſties Name, concerning the buſineſſe of HULL.

LONDON:
Printed by ROBERT BARKER, Printer to the Kings moſt Excellent Majeſtie: And by the Aſſignes of JOHN BILL.
1642.

A pamphlet from 1642, at the outbreak of the English Civil War, in which Charles I responds to another pamphlet issued by Parliament. *The London Library*

monstrance prompted a response entitled "His Majesties Answer to the Petition" from the king, who had by this time left London for the north. Charles expressed his surprise that the Grand Remonstrance was "already abroad in Print" and gave his response, and the two sides went back and forth for several months. Parliament's statements were printed in London and the king's in York, and both were then reprinted elsewhere. This was a dialogue taking place before the public, as was explicitly acknowledged by the title of one of the king's publications: "His Majesties Declaration to both Houses of Parliament; (Which he Likewise Recommends to the Consideration of all His Loving Subjects) In Answer to that Presented to Him at Newmarket the 9th of March 1642." But the king made it clear that he was having this public conversation only reluctantly: "no doubt it will appeare a most triviall and fond Exception, when all Presses are open to vent whatsoever they think fit to say to the people, a thing unwarranted by former custome, that We should not make use of all lawfull meanes to publish Our Just and necessary Answere thereunto." In other words, Charles was using the press in this way only because everybody else was. This was a remarkable change from the tight controls on political material at the start of the century. The public nature of the debate meant that others could join in. Henry Parker, a barrister who supported Parliament, wrote several influential pamphlets providing commentary on the king's replies, such as "Observations upon some of His Majesties late answers and expresses" (1642). These prompted responses from other writers, to which Parker responded in turn. The exchange culminated in June 1642 with Parliament's "Nineteen Propositions," which proposed a new division of power between Parliament and the king, with a much larger role for Parliament. Charles rejected it, and England descended into civil war.

It was a conflict fought in the media as well as on the battlefield. Writing in the 1660s, one chronicler, Richard Atkyns, lamented that with the collapse of press controls, printers "fill'd the Kingdom with so many Books, and the Brains of the People with so many contrary Opinions, that these Paper-pellets became as dangerous as Bullets." Supporters of both king and Parliament took advantage of the freedom of an uncensored media environment to justify their actions, attack their enemies, and appeal to public opinion. During 1642 a throng of weekly newsbooks appeared in London, and in 1643 an official Royalist newsbook, the *Mercurius Aulicus*, was founded in

Oxford, where Charles had established his base. This prompted the establishment of *Mercurius Britannicus*, a parliamentarian rival. Alongside these newsbooks were proclamations, pamphlets, books, ballads, and broadsides in unprecedented numbers, providing propaganda, news, spin, analysis, and commentary.

The average number of titles published in print in England during the 1630s was 624 a year, but the figure jumped to more than two thousand in 1641 and more than four thousand in 1642. In all, some forty thousand different titles appeared between 1640 and 1660. Assuming an average print run of one thousand, that amounts to forty million copies, at a time when the population of England was about five million. This outbreak of printed material put even Luther's campaign in the shade. Other European crises since his time had produced flurries of pamphlets, but in smaller volumes. Around six hundred political pamphlets appeared during the Netherlands Rebellion from 1566 to 1584, for example, and political turmoil in France in 1614–17 resulted in around twelve hundred pamphlets. The momentous shift taking place in England was clear at the time to George Thomason, a London bookseller, who began collecting printed items in 1641. By 1662 he had amassed more than twenty-two thousand items, equivalent to more than half the known publications over the previous two decades. Today his collection, in the British Library in London, provides an invaluable record of the media explosion of the 1640s.

Freedom of the press increased both the volume of titles and their variety. A range of new styles, voices, and uses for printed material emerged, building on the innovations of the previous sixty years, which in retrospect can be seen to have begun with the stylistic and textual trickery of the Marprelate tracts of the 1580s. In the subsequent decades the formal, Ciceronian style of prose writing gradually gave way to a more conversational, vernacular style. News, history, and opinion were freely mixed and presented in a range of formats. Authors of pamphlets worked out how to address an anonymous audience, doing their best to anticipate objections to their arguments and respond to them in advance. Most strikingly of all, the pamphlets of the 1640s existed in an interconnected web, constantly referring to, quoting, or in dialogue with each other, like blog posts today. Very often the jumping-off point of a pamphlet was to respond to, add to, criticize, or praise another pamphlet. Consider, for example, a pamphlet, published in 1646, entitled "XII Resolves Concerning the

Disposall of the Person of the King: in a Sharpre Reproofe to a Rejoynder to Three Pamphlets, Published in Defence of Mr Chaloners Speech (Called, A Speech Without Doores, and Said to be Defended Without Reason) Under Pretence of the Vindication of the Parliaments Honour." It is an answer to a response to three pamphlets that were themselves criticizing a speech, and is therefore four removes from the original text. Such lengthy titles, appearing on the front pages of pamphlets, enabled browsers at bookstands to pick up and follow the thread of the discussions going on among multiple authors, and to position new texts in relation to those they had read before.

Pamphlets could engage with or refer to others in a variety of ways. Some were written as letters, a logical evolution of the scribal form and an obvious way to present an eyewitness account or a polemic, which could be (or pretend to be) addressed to a particular recipient. Such letter pamphlets had titles like "The Intentions of the Army Discovered in a Letter From a Gentleman Residing there to a Friend of His in London" or "A Letter from a Gentleman in the City to a Friend in the Country." They often started by pretending to respond to a previous letter that did not, in fact, exist. Letter pamphlets might then inspire responses in the same format. A more directly adversarial approach was to quote chunks of an opponent's text, either in italics or quotation marks, and to respond to each chunk in turn, a technique that has since been revived by modern bloggers. Further comments on a quoted text might also appear in the margins. When Archbishop Laud was executed in 1645, for example, *Mercurius Britannicus* reprinted and dissected his scaffold speech, many accounts of which were also published separately. Selectively quoting another text in this way imposed a particular reading on it, and could also be used to undermine a rival pamphlet commercially by saving readers from having to buy the original. However, this approach could also unwittingly promote rival arguments by making them more widespread, particularly if the rebuttal was unconvincing.

Responding to another text could also be done by reprinting it in its entirety, followed by a response; reprinting it with criticism interspersed in the body of the text in italics, as though constantly interrupting someone giving a speech; printing passages from an opponent's work and juxtaposing them with contradictory statements from the same author to highlight the inconsistency; or printing extracts from a speech or pamphlet followed by a fictitious dialogue between real or imaginary characters, such as Peace and Truth, discussing its merits or failings, or delivering lengthy speeches of their own. The head (mounted on a pole) of Thomas Bensted, who was executed for treason in 1641, appeared in a pamphlet that year discussing the political situation with another disembodied head. Such "play pamphlets" combined news with satire and were a particularly versatile form.

The bewildering variety of new voices and formats made it very difficult to work out what was going on. As one observer put it in November 1641, "ofttimes we have much more printed than is true." Another writer complained in 1642 that "we are daily eye-witnesses of seditious Pamphlets, which (as false alarms) either terrifie us at first view, with conceited flashes, or else house us in dangerous security by their silver speeches, so that muffled with such penny stuffe, wee cannot judge of affairs without truths enemy, Partiality." As well as confusing the people of the time, this would also confuse future readers. Thomas Fuller, a clergyman and historian, complained that pamphlets "cast dirt on the faces of many innocent persons, which dried on by continuance of time can never be washed off . . . The pamphlets of this age may pass for records with the next (because publicly uncontrolled) and what we laugh at, our children may believe." John Rushworth, who wrote a history of the civil war in 1659 as an antidote to the misrepresentations in the press, observed that "men's fancies were more busy than their hands . . . printing declarations, which were never passed; relating battles which were never fought, and victories which were never obtained; dispersing letters, which were never writ by the authors . . . the impossibility for any man in after-ages to ground a true history, by relying on the printed pamphlets in our days, which passed the press whilst it was without control, obliged me . . . whilst things were fresh in memory, to separate truth from falsehood."

A growing chorus of complaint about the "scurrilous and fictitious pamplets" that printed "rumours mixt with falsity and scandalism," as one tract put it, prompted Parliament to reintroduce licensing in 1643. It had been heavily lobbied by the Worshipful Company of Stationers, which wanted to reclaim its special privileges and its role in regulating the press. Parliament passed an ordinance complaining that "very many . . . have taken upon them to set up sundry private Printing Presses in Corners, and to print, vend, publish and disperse, Books, Pamphlets, and Papers." Rather than requiring approval from bishops, all new works were to be approved by officials appointed by Parliament before being entered into the company's register and printed. Unlicensed presses were outlawed. But the new rules proved to be ineffective in the chaotic environment of the civil war, in which large parts of the country were beyond Parliament's control, and there were more pressing military matters to deal with than press censorship. Complaints about the excesses of the press continued, and the cacophony of an uncontrolled media environment ultimately lasted almost a decade.

## JOHN MILTON AND *AREOPAGITICA*

The breakdown of press regulation was not universally regarded as a bad thing. Indeed, the attempt to reintroduce licensing in 1643 prompted one of the earliest and most eloquent defenses of the principle of freedom of expression: John Milton's *Areopagitica.*

Milton was a brilliant scholar who spent the years after graduating from Cambridge studying, traveling, teaching, and establishing a reputation as a poet. When the political crisis erupted in 1639 he turned from poetry to pamphleteering, writing five pamphlets attacking the bishops of the Church of England for their corruption and intolerance of unorthodox religious views. His next pamphlet, arguing for the legalization of divorce, got him into trouble. He wrote it after his wife, who was seventeen years his junior and whom he had married on a whim, left him after a few weeks and returned to her parents. Milton argued that marriage was meant to be a spiritual bond and that temperamental incompatibility therefore ought to be grounds for divorce. He tried to obtain a license for his divorce pamphlet, in accordance with the new rules introduced in 1643, but was refused. He went ahead and published it anyway, and was then condemned in Parliament as a dangerous radical who was in favor of polygamy. Incensed, Milton responded by attacking the licensing system itself in another pamphlet, "*Areopagitica*: A speech of Mr John Milton for the Liberty of Unlicenc'd Printing," which appeared in November 1644.

*Areopagitica* is notable because it sliced through the existing arguments about what should be allowed to be printed, and who should decide. Milton argued that because humans were fallible, nobody could be a perfectly fair judge of what it was permissible to print. It was far better, he insisted, to allow everything to be published, exposing readers to a range of opinions and letting them make up their own minds: "And though all the winds of doctrine were let loose to play upon the earth, so Truth be in the field, we do injuriously by licensing and prohibiting, to misdoubt her strength. Let her and Falsehood grapple; who ever knew Truth put to the worse, in a free and open encounter?" This view that "truth will out" might be considered naive, but Milton's argument was that it is even more naive to assume that anyone is capable of acting as a reasonable censor. Licensers were susceptible to human error or partisan bias, he noted, and only the "ignorant, imperious, and remiss, or basely pecuniary" would want "to be made the perpetual reader of unchosen books and pamphlets." If someone published something inappropriate, the law could still be brought to bear on them afterward, whether for libel, treason, slander, or blasphemy.

Admittedly, press freedom would allow bad or erroneous works to be printed. But that, Milton argued, was actually a good thing. If more readers came into contact with bad ideas because of printing, those ideas could be more swiftly and easily disproved. He likened truth to a warrior who, rather than having to be cloistered and protected, stays in trim through constant battles with her adversary. This was a Puritan idea, that good emerges only through conflict with and rejection of evil, or what Milton called "knowing good by evil." He called incorrect views "dust and cinders" that "may yet serve to polish and brighten the armory of truth." Being exposed to a wide range of ideas, he argued, strengthened the character of a reader. "A wise man will make better use of an idle pamphlet, than a fool will do of sacred Scripture," he wrote. Besides, bad ideas would spread anyway, even without printing. And with so many printing presses out in the wild, censorship could never be fully effective; widespread access to the technology of publishing made prepublication censorship impractical. Parliament was unable to suppress the royalist weekly *Mercurius Aulicus*, for example, which was resold and sometimes even reprinted in London by royalist sympathizers ("printed, as the wet sheets can witness, and dispers't among us, for all that licencing can doe"). And suppressing dissident views could backfire by giving those views greater prominence. So it was best simply to allow everything to be published, and let rival views compete on the battlefield of ideas.

*Areopagitica* was written as a speech addressed to Parliament, though this was just a literary device: Milton never actually delivered it as a speech. He used several ruses to try to win his intended audience over to his cause. He provided a brief history of censorship, starting with the tolerance shown in ancient Athens, which he flatteringly compared to the English Parliament whose members he was addressing. (The name *Areopagitica* is a reference to the Areopagus, or rock of Ares, the site of the Athenians' highest court.) Milton explained how the Catholic church had gradually extended its banning of books and controls on the press, initially for the purpose of suppressing heresy and then for any reason at all. In doing so he linked the idea of controls on the press with Catholicism, implying to his largely anti-Catholic audience that press freedom was an inherently Protestant position. In lobbying for the 1643 ordinance, the Worshipful Company of Stationers had even pointed out that Catholics tended to be better at censorship: "We must in this give Papists their due; for as well where the Inquisition dominates, as not, regulation is more strict by far, than it is among Protestants." In other words, having broken free of Rome, England should also remove the restrictions on the press. (Milton made only one exception, which looks strikingly illiberal to modern eyes but would have appealed to the anti-Catholic views of his intended audience. He approved of the suppression of Catholic works, because he considered Catholic doctrine to have been utterly discredited on the battlefield of ideas and to have no role in the quest for truth.)

Milton concluded his argument by highlighting the drawbacks of licensing and its negative impact on learning, politics, and religion. Vigorous argument, he wrote, was good for people and for the country as a whole. He called London "the mansion house of liberty" filled with "pens and heads . . . revolving new notions and ideas . . . reading, trying all things, assenting to the force of reason and convincement." He had in mind new religious, political, and even scientific ideas, and his point was that they should all be allowed to be tested by argument. Licensing, by contrast, suppressed the emergence of new ideas, some of which would be valuable, whether or not they were true. It was patronizing, he argued, for a government to look down on its citizens, not daring to "trust them with an English pamphlet"; it would discourage people from thinking for themselves if they relied on the government to tell them what was right. A vigorous, healthy society, he concluded, required that vibrant, free-wheeling discussion should not merely be tolerated, but encouraged.

To emphasize his opposition to licensing, Milton published his speech as an unlicensed pamphlet. Its form also underlined his argument in another way: for although few people could address Parliament in person, anyone could, in theory, publish a pamphlet taking the form of a speech to Parliament (or, indeed, to any other notional audience). *Areopagitica* illustrated, in both form and argument, that a free press had the merit of creating a much broader and more open public space for debate and discussion—one that was not limited to members of Parliament.

Although Milton's expression of these views is best known today, he was not alone in holding or espousing them. The idea of press freedom was in the air at the time, particularly among proponents of religious tolerance. Having thrown off the heavy hand of royal regulation, pamphleteers realized that parliamentary regulation would be just as bad, and argued that it was better to have no prepublication regulation at all. Henry Robinson, a merchant and writer who, like Milton, was an advocate of religious tolerance, also argued that simply allowing everything to be published, without trying to vet it first, was the best way to discover and spread the truth. In "Liberty of Conscience: Or the Sole Means to Peace," published shortly before *Areopagitica* in 1644, he argued that "it were better that many false doctrines were published, especially with a good intention and out of weaknesse only, than that one sound truth should be forcibly smothered or wilfully concealed; and by the incongruities and absurdities which accompany erroneous and unsound doctrines, the truth apears still more glorious, and wins others to the love thereof." Just as Milton called for letting truth and falsehood grapple, Robinson asked: "Doe we suspect that errour should vanquish truth?" He went even farther than Milton, calling for freedom of expression in all forms of media, even those that had not yet been invented. The fight between truth and falsehood, he declared, "must be fought out on eaven ground, on equall termes, neither side must expect to have a greater liberty of speech, writing, Printing, or whatever else, than the other."

In the short term, campaigners for freedom of expression got what they wanted, because the 1643 licensing rules were so feebly enforced. Only after the defeat and execution of Charles I was effective control of the press reestablished by Oliver Cromwell, who took charge as the republican head of state in 1653. But in the decades that followed there were repeated outbursts of unlicensed printing and public debate whenever the political situation became unstable and control of the press weakened. In the uncertainty that followed Cromwell's death in 1658 there was a sudden outpouring of polemical pamphlets and newsbooks (in June 1659 Edward Nicholas wrote of Cromwell that "the monster is now understood by every Pamphleteer"). Control was once again established in 1662 after the restoration two years earlier of the monarchy under the new king, Charles II. During his reign control of the press was assigned to Roger L'Estrange, a royalist pamphleteer who believed that allowing the unrestricted circulation of news pamphlets was dangerous because it made "the multitude too familiar with the actions and counsels of their superiors." He became known as the "bloodhound of the press" for his ability to sniff out clandestine printing presses and unlicensed works. He regarded anyone involved in spreading a text as responsible for its contents: authors, printers, publishers, "Hackny-Coachmen, Carryers, Boat-men." He proposed not just prepublication licensing, but checks at every stage of the process, and even had one man executed for treason merely because he had proofread an offending pamphlet.

L'Estrange held this position until the 1662 rules lapsed in 1679. There followed a further outburst of unlicensed printing associated with the controversy over the "Popish plot," a fabricated conspiracy intended to discredit Charles II's Catholic brother, James, and prevent him from becoming the next king. Hundreds of pamphlets made claims and counter-claims, and the debate over press controls was reopened, with Milton's words dusted off and recycled by many pamphleteers. It felt like a return to the freedom of the 1640s, as many noted at the time. But James eventually became king in 1685 and reintroduced licensing. When he was ousted in 1688 in the Glorious Revolution, in favor of his Protestant daughter Mary and son-in-law William, there was another outbreak of printing, but it was soon brought under tight control again. Then, in the early 1690s, a squabble broke out over the redrafting of the rules concerning press regulation. So many objections were raised to small details of the act that Parliament ran out of time, and the existing act expired in 1695. By accident rather than design, the regime of prepublication licensing had finally come to an end. People could print whatever they wanted (though authors and publishers could still be prosecuted for treason, blasphemy, and libel after publication). A century and a half of control, of varying degrees of effectiveness, had given way to a free press.

The struggle for press freedom in seventeenth-century England, and the philosophical arguments used to advocate it, went on to inspire both French and American revolutionary thought during the eighteenth century. In France, the Comte de Mirabeau translated *Areopagitica* and declared that its subject matter was "suitable for all times and all countries." Milton was ultimately named one of the "founding fathers of the French republic" by Jacques-Pierre Brissot, one of the leaders of the French Revolution. In America, Milton's influence was clearly visible in Benjamin Franklin's "Apology for Printers" (1731): "Printers are educated in the belief, that when men differ in opinion, both sides ought equally to have the advantage of being heard by the public; and that when truth and error have fair play, the former is always an overmatch for the latter." Milton laid down some of the intellectual foundations of the First Amendment to the U.S. Constitution, and *Areopagitica* has been cited many times in rulings by the United States Supreme Court, invariably to support a broad protection of freedom of speech or publication. More than three hundred years after its publication, *Areopagitica* was being cited in arguments against restrictions on information about contraception, for example.

The ethos of the 1640s, it seems, is alive and well today. For the first time, politics became a public conversation, a discussion driven by the sharing of media. As well as making it quicker and easier to spread political ideas, freer printing changed the nature of political discourse as participants sought to appeal to and influence public opinion. Disputes between rival pamphleteers, and the ability to compare documents from rival points of view, encouraged people to consider the political process as a vast national discussion. One outcome was the modern notion of freedom of the press; another was the emergence of the first publications to cover domestic politics without restraint, which led to the earliest modern newspapers, though they took a few decades to evolve into that form.

To modern eyes, the chaotic and adversarial media environment of the 1640s has much in common with the Internet's blogging culture. The blogosphere, like the pamphlets of the English Civil War period, is an interconnected web of documents that refer to and argue with each other, written by occasionally anonymous, sometimes unreliable, and often fiercely partisan authors who are free to publish whatever they like, and whose posts can assume a wide range of formats. Some modern bloggers, recognizing this parallel, have even adopted seventeenth-century pseudonyms. This seems only appropriate. Today we think nothing of the ability to publish material online with a few clicks, or to read a wide variety of conflicting and unfiltered opinions. It is a freedom that has its roots in a fertile decade during which social media flowered in England, nearly four hundred years ago.

# CHAPTER 6

# AND SO TO THE COFFEEHOUSE: HOW SOCIAL MEDIA PROMOTES INNOVATION

*Where can young gentlemen, or shop-keepers, more innocently and advantageously spend an hour or two in the evening, than at a coffeehouse? Where they shall be sure to meet company, and, by the custom of the house, not such as at other places, stingy and reserved to themselves, but free and communicative.*

—from "Coffee-Houses Vindicated" (1675)

## SOCIAL NETWORKING BY THE CUP

IN THE MIDDLE of the seventeenth century an innovation from the Arab world transformed the sharing and distribution of information in western Europe. It provided a new environment in which people from a variety of backgrounds could find others with similar interests, and then read and discuss a wide range of media in a congenial setting. Not everyone welcomed the freedom of speech afforded by this new social forum, and some people worried that its compelling, information-rich environment, which provided an endless and addictive stream of trivia, gossip, and falsehood, was distracting people from more productive pursuits. Others, however, approved wholeheartedly of this egalitarian new intellectual space—the coffeehouse—and believed that it offered "the sanctuary of health, the nursery of temperance, the delight of frugality, an academy of civility, and a free-school of ingenuity."

Coffee had become popular in the Arab world in the late fifteenth century and was the first caffeinated drink to reach Europe, in around 1600. Its mentally stimulating effect ensured that it quickly moved beyond its early popularity among European botanists and medical men. And along with the coffee bean itself came the institution of the coffeehouse, which had become an important meeting place and source of news in the Arab world. William Biddulph, an English traveler, had noted in 1609 that "their Coffa houses are more common than Ale-houses in England . . . if there be any news it is talked of there." The first European coffeehouses opened in Venice in the 1640s, Britain in the 1650s, and the Netherlands in the 1660s. In Europe, and particularly in Puritan England, coffeehouses were embraced as a more respectable alternative to taverns. Instead of serving alcohol, which dulled the mind, coffeehouses served coffee, which sharpened it. Unlike taverns, which were often somewhat shady establishments, coffeehouses were well lit and adorned with bookshelves, mirrors, pictures in gilt frames, and good furniture. Coffee drinking was considered a sign of sophistication and openness to new ideas. A pamphleteer in 1672 observed that coffee had so much "credit got . . . he's no gentleman that drinks it not." It swiftly became the preferred drink of scientists, intellectuals, merchants, and clerks, and coffeehouses established themselves as centers of information exchange where the latest pamphlets, broadsheets, gazettes, and newsletters could be read and discussed. As a popular rhyme, "News from the Coffee-House," put it in 1667:

*You that delight in Wit and Mirth, and long to hear such News,*
*As comes from all parts of the Earth, Dutch, Danes, and Turks and Jews,*
*I'le send you a Rendezvous, where it is smoaking new:*
*Go hear it at a Coffee-house—it cannot but be true . . .*
*There's nothing done in all the World, From Monarch to the Mouse,*
*But every Day or Night 'tis hurl'd into the Coffee-house.*

Of all Europe's cities, London embraced the coffeehouse most quickly and wholeheartedly. The city's first coffeehouse was opened in 1652 by Pasqua Rosée, the Armenian servant of an English merchant who had acquired a taste for coffee while traveling in the Middle East. It was an instant success, prompting local tavern keepers to protest to the Lord Mayor that Rosée had no right to set up a business in competition with them, since he was not a freeman of the City. Rosée was ultimately forced out of the country, but the idea of the coffeehouse had taken hold. By 1663, the number of coffeehouses in London had reached eighty-three. Many were destroyed in the Great Fire of London in 1666, but even more arose in their place, and by the end of the century there were around five hundred and fifty of them in the capital alone. Noting that the planet Saturn takes twenty-nine years to complete one orbit of the sun, one pamphleteer remarked in 1675 that "Saturn has not finished one revolution through the orb since coffee-houses were first known amongst us, yet 'tis worth our wonder to observe how numerous they are already grown, not only here in our metropolis, but in both universities and most cities and eminent towns throughout the nation." It was a similar tale in France. After the first coffeehouse in Paris, Café Procope, opened its doors in 1686, the number of coffeehouses in the city quickly mushroomed, reaching three hundred eighty by 1720, six hundred by 1750, and eight hundred by 1800.

Coffeehouses in large cities often specialized in the discussion of particular subjects, depending on the dominant activity in their neighborhood. In London, those around Saint James's and Westminster were frequented by politicians, and those near Saint Paul's Cathedral by clergyman and theologians. The literary set, meanwhile, congregated at Will's coffeehouse in Covent Garden, where for three decades the poet John Dryden and his circle reviewed and discussed the latest poems and plays. Coffeehouses around the Royal Exchange were thronged with businessmen, who would keep regular hours at particular coffeehouses so that their associates would know where to find them, and who used coffeehouses as offices, meeting rooms, and venues for trade. In Jonathan's, in Exchange Alley, customers bought and sold stocks and commodities, the prices of which were posted on the wall. Merchants and shipowners met in Lloyd's coffeehouse. Books were sold at Man's coffeehouse in Chancery Lane, auctions took place at the London, doctors went to Baston's, scientists to the Grecian, and barristers to the George. Similarly, in Paris, poets met at Café Procope and Café Parnasse, literati at the Café Bourette, actors at the Café Anglais, musicians at Café Alexandre, and army officers at the Café des Armes d'Espagne. The Café des Aveugles doubled as a brothel.

Some people frequented multiple coffeehouses, the choice of which depended on their interests. A merchant, for example, might divide his day between a financial coffeehouse and one specializing in Baltic, West Indian, or East Indian shipping. And one way to quantify the wide-ranging interests of the English scientist Robert Hooke is to note that according to his diary, he frequented around sixty London coffeehouses during the 1670s. So closely were some coffeehouses associated with particular topics that the *Tatler*, a London magazine founded in 1709, used the names of coffeehouses as subject headings for its articles. Its first issue declared:

> All accounts of Gallantry, Pleasure, and Entertainment shall be under the Article of White's Chocolate-house; Poetry, under that of Will's Coffee-house; Learning, under the title of Grecian; Foreign and Domestick News, you will have from St James's Coffee-house.

Whatever the topic, the main business of coffeehouses was the sharing and discussion of news and opinion in spoken, written, and printed form; their patrons wanted to imbibe information as well as coffee and tobacco. Entering a coffeehouse, one would be greeted through thick clouds of tobacco smoke by the cry "What news have you?" and would then look for a space at a large table covered with papers of various kinds. One writer in 1682 explained that coffeehouse patrons might expect to find "a table of an acre long covered with nothing but tobacco-pipes and pamphlets." Coffeehouses subscribed to periodicals and gathered together a wide range of material; some were supplied with foreign journals and gazettes, or subscribed to manuscript newsletters. An account from 1707 remarked that "the Coffeehouses particularly are very commodious for a free Conversation, and for reading at an easie Rate all manner of printed News, the Votes of Parliament when sitting, and other Prints that come out Weekly or casually."

After taking a seat and paying a penny one would be served a shallow cup, or dish, of coffee, and could then spend as long as one liked reading and discussing the latest news and pamphlets with other patrons. Sometimes a text would be read aloud by one person at a table, with pauses for explanation and discussion. Conversation between strangers was encouraged, and distinctions of class and status were to be left at the door. One set of printed coffeehouse rules stipulated that "no man of any station need give his place to a finer man," and a seventeenth-century rhyme explained that "gentry, tradesmen, all are welcome hither, and may without affront sit down together." The satirist Samuel Butler wrote that a coffee-seller "admits of no distinction of persons, but gentleman, mechanic, lord, and scoundrel mix, and are all of a piece." In theory, at least, this was a realm of pure information exchange, where ideas were to be scrutinized, combined, or discarded on their own merits, and people could speak their minds.

Such was the popularity of coffeehouses, and the freedom of speech customary within them, that Charles II tried to close them down in 1675, fearing that they were centers of seditious plotting. A royal proclamation declared that coffeehouses had produced "very evil and dangerous effects . . . for that in such Houses . . . divers False, Malitious and Scandalous Reports are devised and spread abroad, to the Defamation of His Majestie's Government, and to the Disturbance of the Peace and Quiet of the Realm." The result was an instant public outcry, for coffeehouses had by this time become central to London's commercial and political life. Coffeehouse proprietors led by Thomas Garraway organized a petition in protest. The chancellor of the exchequer took their side, reminding the king that the sale of coffee, tea, and chocolate in coffeehouses produced valuable tax revenues. And some of the king's advisers expressed doubts about the legality of the ban. The king swiftly backtracked. The proclamation, issued on December 29, 1675, was withdrawn on January 8, 1676. Instead, it was announced that each coffee-seller would be allowed to stay in business for six months if he paid £500, swore an oath of allegiance, and did his best "to prevent and hinder all Scandalous Papers, Books or Libels concerning the Government, or the Publick Ministers thereof, from being brought into his House, or there to be Read, Perus'd or Divulg'd." But the fee and time limit were soon dropped, along with the impractical requirement that proprietors had to police the content being shared within their walls, and everyone simply carried on as before. Coffeehouses had successfully pushed back against regulation and defended their valued status as forums for free speech and the free exchange of ideas.

The effect of these caffeine-powered hubs was to increase the speed and efficiency with which information percolated through society. Coffeehouses imposed order on the chaotic media environment of the time, sorting material by topic and making it much easier to find specific types of information, and people to discuss it with. Both pamphlets and people, to use the modern term, became more "discoverable." Coffeehouses gave physical form to the previously immaterial social networks along which information passed, making it much easier to connect to them. If you wanted to know what London's scientists were talking about, for example, and make contact with them, all you had to do was walk into the Grecian coffeehouse. The social mixing that took place in coffeehouses allowed ideas to leap over the boundaries of England's class system, as the writer John Aubrey observed when he praised the "modern advantage of coffeehouses . . . before which, men knew not how to be acquainted, but with their own relations, or societies." Anyone launching a new poem or pamphlet could leave copies in the establishments where it was most likely to find a receptive audience. And pamphleteers of every stripe, notably Jonathan Swift and Daniel Defoe, found that coffeehouse discussions offered a rich source of material for their sharp-witted satires. Coffeehouses became the logical place not just to read new works, but to write them, too. After the final collapse of press regulation in 1695 a host of new periodicals appeared, including Ned Ward's *London Spy*, Defoe's *Review*, Swift's *Examiner*, Richard Steele's *Tatler*, and the *Spectator*, run by Steele and Joseph Addison. Some establishments even started issuing their own specialist newsletters to cater to their patrons.

A further boost to the flow of information came from the use of coffeehouses as mailing addresses, which became commonplace after the establishment of the London penny-post in 1680. Street numbering had yet to be introduced, and coffeehouses had already established themselves as information hubs, so they were the logical places to send and receive mail. An account from the early eighteenth century describes how this worked: "To receive Letters and Parcels to be sent away, there are near 500 Shops and Coffee-Houses in the City and Country, where the Messengers having their respective Walks, collect them, and carry 'em to the proper Office. In most places of London and Westminster, they do it every Hour of the Day; and in the remotest parts, every two Hours. In the Towns near London twice a Day; and in remoter Parts but once." Regulars at a particular coffeehouse would pop in once or twice a day, drink a dish of coffee, hear the latest news, and check to see if there was any new mail waiting for them. With their promise of a constant and unpredictable stream of news, messages, and gossip, coffeehouses were an alluring social platform for sharing information.

The interior of a London coffeehouse, where people went to read, discuss, and (sometimes) write pamphlets. *Mary Evans Picture Library*

## ARE COFFEEHOUSES MAKING US STUPID?

So seductive was this new environment—one never knew what one might learn on one's next visit, or whom one might meet—that coffeehouse denizens found themselves whiling away hours in reading and discussion, oblivious to the passage of time. "Thence to the coffeehouse" appears frequently in the celebrated diary of Samuel Pepys, an English public official. His entry for January 11, 1664, gives a flavor of the cosmopolitan, serendipitous atmosphere that prevailed within the coffeehouses of the period, where matters both trivial and profound were discussed.

> Thence to the Coffee-house, whither comes Sir W. Petty and Captain Grant, and we fell in talke (besides a young gentleman, I suppose a merchant, his name Mr Hill, that has travelled and I perceive is a master in most sorts of musique and other things) of musique; the universal character; art of memory . . . and other most excellent discourses to my great content, having not been in so good company a great while, and had I time I should covet the acquaintance of that Mr Hill . . . The general talke of the towne still is of Collonell Turner, about the robbery; who, it is thought, will be hanged.

Enthusiasm for coffeehouses was not universal, however, and some observers regarded them as a worrying development. They grumbled that Christians had taken to a Muslim drink instead of traditional English beer, and fretted that the livelihoods of tavern-keepers might be threatened. But most of all they lamented, like critics of social media today, that coffeehouses were distracting people and encouraging them to waste time sharing trivia with their friends when they ought to be doing useful work.

When coffee became popular in Oxford and coffeehouses began to multiply, the university authorities objected, fearing that this was promoting idleness and diverting students from their studies. Anthony Wood, an Oxford antiquarian, was among those who denounced the enthusiasm for the new establishments. "Why doth solid and serious learning decline, and few or none follow it now in the university?" he asked. "Answer: Because of coffee-houses, where they spend all their time." Similar concerns were voiced in Cambridge, where one observer noted that it "is become a custom after chapel to repair to one or other of the coffee houses (for there are divers), where hours are spent in talking, and less profitable reading of newspapers, of which swarms are continually supplied from London. And the scholars are so greedy after news (which is none of their business) that they neglect all for it, and it is become very rare for any of them to go directly to his chamber after prayers without first doing his suit at the coffee-house, which is a vast loss of time grown out of a pure novelty. For who can apply close to a subject with his head full of the din of a coffee-house?"

Inevitably, the opposition to coffeehouses found expression in pamphlet form. The author of "The Grand Concern of England Explained" (1673) grumbled that coffeehouses had

> done great mischiefs to the nation, and undone many of the King's subjects: for they, being great enemies to diligence and industry, have been the ruin of many serious and hopeful young gentlemen and tradesmen, who, before frequenting these places, were diligent students or shopkeepers, extraordinary husbands of their time as well as money; but since these houses have been set up, under pretence of good husbandry, to avoid spending above one penny or two-pence at a time, have gone to these coffee-houses; where, meeting friends, they have sat talking three or four hours; after which, a fresh acquaintance appearing, and so one after another all day long, hath begotten fresh discourse, so that frequently they have staid five or six hours together in one of them; all which time their studies or shops have been neglected.

The coffeehouse bore, the know-it-all political commentator, and the businessman spreading false rumors are all stock figures in the satire of the period. Another pamphlet, "The Character of a Coffee-house" (1673), mocks the coffeehouse as

> an exchange, where haberdashers of political small-wares meet, and mutually abuse each other, and the publick, with bottomless stories, and headless notions; the rendezvous of idle pamphlets, and persons more idly employed to read them . . . The room stinks of tobacco worse than hell of brimstone, and is as full of smoke as their heads that frequent it.

By contrast, coffeehouse supporters such as Joseph Addison, the founder of the *Spectator*, regarded them as places where young people could go to improve themselves, find out about the world, and learn better manners and good taste. Like the *Tatler* before it, the *Spectator* both commented upon and stimulated coffeehouse discussions. As well as serving as a helpful digest for readers outside London, it created in effect a single idealized coffeehouse in which Addison strove to instill "sound and wholesome sentiments" into his readers. "It was said of Socrates that he brought Philosophy down from heaven, to inhabit among men," Addison wrote in the March 12, 1711, edition of the *Spectator*, "and I shall be ambitious to have it said of me, that I have brought Philosophy out of closets and libraries, schools and colleges, to dwell in clubs and assemblies, at tea-tables and in coffee-houses." The author of the pamphlet "Coffee-houses Vindicated" made another classical reference when extolling coffeehouses' character-improving virtues: "In brief, it is undeniable, that, as you have here the most civil, so it is, generally, the most intelligent society; the frequenting whose converse, and observing their discourses and deportment, cannot but civilize our manners, enlarge our understandings, refine our language, teach us a generous confidence and handsome mode of address, and brush off that *pudor subrusticus* (as, I remember, Tully somewhere calls it), that clownish kind of modesty frequently incident to the best natures, which renders them sheepish and ridiculous in company." (Cicero, referred to here by his nickname "Tully," had talked of his own "clownish bashfulness" in a letter to his friend Lucius Lucceius some seventeen centuries earlier.)

Whether or not they improved people's manners, there was no doubt that coffeehouses were popular venues for educational discussion. Far from discouraging intellectual activity, coffeehouses actively promoted it. Indeed, they were sometimes called "penny universities," because anyone could enter, and then listen to or join in the conversation, for the price of a dish of coffee. As one ditty of the time put it: "So great a Universitie, I think there ne'er was any; In which you may a Scholar be, for spending of a Penny." Highbrow coffeehouse discussions were particularly popular among those who took an interest in the progress of science, or "natural philosophy" as it was known at the time. As Hooke's diary shows, he and his scientific colleagues used coffeehouses as venues for academic debates, negotiations with instrument-makers, and, on occasion, scientific experiments.

## COFFEEHOUSES AND COLLABORATIVE INNOVATION

Hooke and several of his scientific colleagues, including Christopher Wren and Robert Boyle, had acquired a taste for coffee in Oxford during the 1650s, when they had all been members of a club of science enthusiasts formed by John Wilkins, a senior academic at the university. They continued their tradition of coffeehouse discussion in London, where theirs was one of several groups that came together in 1660 to form the Royal Society, Britain's pioneering scientific institution. The society's members, who also included Pepys, Isaac Newton, and Edmond Halley, would often decamp to a coffeehouse after its meetings to continue their discussions. On May 7, 1674, for example, Hooke recorded in his diary that he demonstrated an improved form of astronomical quadrant at a meeting of the Royal Society, and then repeated his demonstration afterward at Garraway's coffeehouse, where he discussed it with John Flamsteed, an astronomer who was appointed by Charles II as the first Astronomer Royal the following year. On another occasion a group of scientists, led by Newton and Halley, dissected a dolphin at the Grecian.

Compared with the formality of the society's meetings, coffeehouses provided a relaxed atmosphere that promoted more freewheeling discussion and speculation. Hooke's diary gives examples of how ideas were tossed around, with information jotted on scraps of paper or scribbled into notebooks. At one meeting, at Man's coffeehouse, Hooke and Wren traded information about the behavior of springs. "Discoursed much about Demonstration of spring motion. He told a pretty thought of his about a poysd weather glass . . . I told him an other . . . I told him my philosophicall spring scales . . . he told me his mechanick rope scale." On another occasion Hooke exchanged recipes for medical remedies with a friend at St. Dunstan's coffeehouse. Such discussions allowed scientists to try out half-formed ideas on their colleagues, suggesting new avenues of investigation and sparking new theories.

Indeed, it was an argument in a coffeehouse that led to the publication of the greatest book of the scientific revolution. On a January evening in 1684, a coffeehouse discussion between Hooke, Halley, and Wren turned to the theory of gravity, the topic of much speculation at the time. Between sips of coffee, Halley wondered aloud whether the elliptical shapes of planetary orbits were consistent with a gravitational force that diminished with the inverse square of distance. Hooke declared that this was the case, and that the inverse-square law alone could account for the movement of the planets, something for which he claimed to have devised a mathematical proof. But Wren, who had tried and failed to produce such a proof himself, was unconvinced and offered a valuable book as a prize if either of his friends could produce a proof within two months.

The book went unclaimed, but a few months later Halley visited Newton in Cambridge. Recalling his coffeehouse discussion with Wren and Hooke, Halley asked Newton the same question: would an inverse-square law of gravity give rise to elliptical orbits? Like Hooke, Newton claimed to have proved this already. Although he could not find the proof when Halley asked to see it, he devoted himself to the problem and sent Halley the proof soon afterward. This proved to be just a foretaste of what was to come. Halley's question had given Newton the impetus he needed to formalize many years of work to produce one of the greatest books in the history of science: *Philosophiae naturalis principia mathematica* ("Mathematical Principles of Natural Philosophy"), generally known as the *Principia*. In this monumental work, published in 1687, Newton demonstrated how his principle of universal gravitation could explain the motions of both earthly and celestial bodies, from the (probably apocryphal) falling apple to the orbits of the planets.

Hooke insisted that he had given Newton the idea of the inverse-square law in letters exchanged a few years earlier. But when he made his case in another coffeehouse discussion after the presentation of the first volume of the *Principia* to the Royal Society in June 1686, Hooke failed to convince his scientific colleagues. There was a world of difference between advancing an idea in the speculative environment of a coffeehouse and proving its correctness. Unlike Newton, Hooke had not published his ideas or formally presented them to the society, and he was always claiming to have thought of other people's ideas first (though, in many cases, he actually had). "Being adjourned to the coffee-house," Halley wrote to Newton, "Mr Hooke did there endeavour to gain belief, that he had such thing by him, and that he gave you the first hint of this invention. But I found that they were all of the opinion, that . . . you ought to be considered as the inventor." The coffeehouse had given its verdict, which still stands today.

The coffeehouse spirit of discussion and innovation was not limited to scientific matters, and extended into the sphere of business and finance. Jonathan's, which was one of Hooke's favorite haunts, was the main coffeehouse in which stock and shares were traded, by virtue of its proximity to the Royal Exchange. Traders and merchants kept particular tables at Jonathan's, from which they did their business. Coffeehouses competed to provide information to their customers, sending boys to the docks to collect gossip and note the arrival and departure of ships. Such news was posted on boards inside coffeehouses or, in some cases, compiled into newsletters. Jonathan's produced its own newsletter, as did one of its regular patrons, Jonathan Castaing, who published "The Course of the Exchange and Other Things" from his "office in Jonathan's Coffee-house" every Tuesday and Friday starting in March 1697. Castaing's newsletter listed exchange rates and the prices of gold, silver, various companies' shares, and various government securities. It was distributed to many coffeehouses, came to be regarded as a trusted source by merchants in London and beyond, and has been issued ever since. (It is now known as the Stock Exchange Daily Official List, a data series that extends right back to 1697.) Jonathan's also gave birth to the London Stock Exchange, after a group of its patrons established a formal stockbroking club in 1761. In 1773 some of them decamped to a new building, initially known as New Jonathan's. But this name did not last long, as the *Gentlemen's Magazine* reported: "New Jonathan's came to the resolution that instead of its being called New Jonathan's, it should be called The Stock Exchange, which is to be wrote over the door." This was the forerunner of the London Stock Exchange.

Around the corner from Jonathan's, in Lombard Street, was Lloyd's coffeehouse, a popular meeting place for ship captains, shipowners, and merchants, who went to hear the latest maritime news and attend auctions of ships and their cargoes. Its proprietor, Edward Lloyd, began to collect and summarize this information, supplemented with reports from a network of foreign correspondents, in the form of a regular newsletter, initially handwritten and later printed and sent to subscribers. Lloyd's became the natural meeting place for shipowners and the underwriters who insured their ships. Some underwriters began to rent regular booths at Lloyd's, and in 1771 a group of seventy-nine of them collectively established the Society of Lloyds, which survives to this day as Lloyd's of London, the world's leading insurance market.

And just as Newton's *Principia*, the foundational work of modern science, had its origins in a coffeehouse, so too did the work that played the equivalent role in economics: *The Wealth of Nations*, by the Scottish economist Adam Smith. He wrote much of his book in the British Coffee House, his base and postal address in London and a popular meeting place for Scottish intellectuals, among whom he circulated chapters of his book for criticism and comment. No doubt there was some time-wasting in coffeehouses, as their critics claimed. But coffeehouses also provided a lively intellectual and social environment in which people could meet and ideas could collide in unexpected ways, producing a stream of innovations that shaped the modern world. On balance, the introduction of coffeehouses did far more good than harm, which should give those concerned about the time-wasting potential of Internet-based social platforms pause for thought. What new ideas and unexpected connections might be brewing in Twitter's global coffeehouse?

## HENRY OLDENBURG ADDED YOU AS A FRIEND

The coffeehouse was not the only new platform for academic discussion to emerge in the mid-seventeenth century. It was complemented by another new forum that also encouraged discussion and collaboration, but this time over long distances: the scientific journal. A key figure in this development was Henry Oldenburg, the secretary of the Royal Society. He was one of several prolific letter-writers, known as "intelligencers," whose far-reaching social connections linked the scientists of Europe together in an information-sharing network during the seventeenth century. Oldenburg professed himself "happy in the acquaintance of many intelligent friends" in foreign parts, by which he meant that he was in regular contact by letter with many of Europe's leading scientists. Born in Germany, he spent several years traveling in Europe as a tutor, making many contacts in the process. He then settled in London, where he was involved with the Royal Society from its inception. His dozens of correspondents included Christiaan Huygens, the Dutch scientist who identified the rings of Saturn and invented the pendulum clock; Johannes Hevelius, the Polish astronomer renowned for mapping the Moon; the eminent Dutch philosopher Benedictus de Spinoza; and Robert Boyle, based in Oxford, who relied on Oldenburg to keep him informed of the latest scientific goings-on in London.

Oldenburg acted as a one-man clearinghouse for scientific news, passing on papers and letters containing newsworthy items from his correspondents to be read out at Royal Society meetings, and writing back to them with summaries of the latest developments both in London and across Europe. A typical letter from Oldenburg to Boyle in 1664 summarizes a Royal Society meeting at which Dr. Walter Pope's letter from Italy, describing terrible storms, had been read out; plans had been made to set up a giant pendulum in Saint Paul's Cathedral; and the nature of poisonous gases in coal mines had been discussed. In this way Oldenburg allowed people who were not able to attend the society's meetings in person to participate in its discussions at a distance. By gathering and redistributing information, Oldenburg and other, earlier intelligencers—such as Samuel Hartlib in England, and Nicolas-Claude Fabri de Peiresc and Marin Mersenne in France—enabled their networks of contacts to work together in informal academies, or so-called "invisible colleges."

But this involved writing everything out by hand. In 1662 one of Oldenburg's correspondents wrote to him that "I have long mervayled, that you never condescend to use the Mercuryes for communications. Tis the expedient of late dayes." Just as printed newsbooks, or mercuries, could be sent in a letter instead of writing out lots of news by hand, could the same not be done for scientific correspondence? Further inspiration came in a letter from a correspondent in Paris, who informed Oldenburg of a plan to publish a printed journal of "all what passeth in Europe in matters of knowledge both philosophical and political," including book reviews, obituaries of notable people, news from libraries and universities, accounts of new scientific discoveries and artistic achievements, and noteworthy legal rulings. Given his excellent connections, Oldenburg had been asked to contribute to this new journal as its correspondent in England, in return for receiving a copy of each issue. This astonishingly ambitious publication was the *Journal des Sçavans* (the "Journal for Learned People"), the first issue of which appeared in January 1665. It inspired Oldenburg to do something similar, but focused specifically on scientific endeavors. His aim, he explained to one of his correspondents, was "to inform the curious what passeth up and downe in the World in matter of Knowledge and Philosophy."

A printed journal rounding up the most recent discussions at the Royal Society, and other scientific news that had been reported to Oldenburg by letter, made sense for several reasons. For one thing, it would be a more efficient way for Oldenburg to disseminate news of the latest scientific goings-on to people outside London. He was struggling to maintain a detailed written correspondence with dozens of scientists; printing his summaries of meetings, and letters and papers in full, so they could be more widely distributed like a newsbook, would be much simpler. In the aftermath of the English Civil War, printed material still had a reputation in England for being scurrilous and unreliable, it was true, but the Royal Society's royal charter gave it unusual credibility. People who could not attend Royal Society meetings would be able to read about what had happened, build upon the results of others, and then send in their own papers to be read out at the meetings and published in the journal. As well as broadening participation in the society's distributed community, the new journal would provide a permanent record of the society's activities and, as a consequence, of the progress of scientific knowledge.

Crucially, from the point of view of participants in the discussion that took place within its pages, the journal would also act as a neutral public forum in which scientists could claim priority for their work. As Oldenburg put it in a letter to Boyle, "jealousy about the first authors of experiments . . . is not groundless; and [I] therefore offer myself to register all those you, or any person, shall please to communicate as new." Writing lengthy scientific books was seen as respectable, but claiming priority for a new discovery by publishing details in a printed pamphlet was considered to be vulgar and boastful. Sending a letter to the Royal Society for publication in its journal, however, provided a more modest and acceptable way to ensure that the news reached a wide audience, and was far more efficient than writing to lots of people individually. And because the journal would appear under the society's imprimatur, Oldenburg asked members of its council to vet the contents of each issue—the origin of the practice of "peer review." This meant that published papers had the society's explicit endorsement. A further attraction of a regular journal was that it would be able to alert its readers to new books, and provide summaries and reviews of them. The practice of providing information about books, in the form of bibliographies, summaries, and reviews, was (along with the filing of useful snippets in miscellanies or commonplace books) one of the tactics intellectuals developed to deal with the feeling of information overload caused by the advent of printing.

The first issue of Oldenburg's journal, the *Philosophical Transactions of the Royal Society of London*, appeared in March 1665. It is a clear descendant of the newsbook, being made up largely of letters or summaries of letters by multiple authors, and with scientific observations of great importance (Hooke's discovery of Jupiter's Great Red Spot) appearing alongside more familiar newsbook fare (a monstrous calf born in Hampshire). In a brief introduction, Oldenburg explained that he thought it "fit to employ the Press, as the most proper way to gratify" people who wished to keep up with the latest scientific developments. This would ensure that "those, addicted to and conversant in such matters, may be invited and encouraged to search, try, and find out new things, impart their knowledge to one another, and contribute what they can to the Grand design of improving Natural knowledge, and perfecting all Philosophical Arts, and Sciences." The first issue also contained an outline of Boyle's new book "An Experimental History of Cold," and an obituary of the French mathematician Pierre de Fermat that included a list of his most important works with brief summaries of their contents.

Thereafter the journal appeared monthly. Around 1,250 copies of each issue were printed; of these, fifty went to Oldenburg to send to his various correspondents, and the rest were distributed for sale in bookshops in England and on the continent. Although the journal was published with the approval of the Royal Society, the perpetually cash-strapped Oldenburg was allowed to run it as a private commercial venture, and he hoped it would provide him with a steady income. But after just five issues an outbreak of the plague in London in July 1665 held up the printing and distribution of the new journal. The Great Fire of London the following year caused further difficulties, because unsold copies had been stored in Saint Paul's Cathedral for safety, along with many books, and all were destroyed when it burned down. Oldenburg also became embroiled in a seemingly endless series of disputes with printers about pricing and distribution, and was repeatedly forced to reduce his share of the sales revenue in order to keep the journal afloat.

Things got even worse in 1667, when Oldenburg was jailed on suspicion of espionage. The writer John Evelyn noted in his diary that

Oldenburg had been "put into the Tower, for writing newes to a virtuoso in France, with whom he constantly corresponds in philosophical matters." The political situation in Europe at the time was very delicate, with England, France, Spain, Sweden, and the Dutch Republic embroiled in the so-called War of Devolution. But scientists considered themselves above such nationalistic squabbles, and Oldenburg had won a special dispensation that allowed him to maintain his correspondence with scientists in Europe—on the condition that he allowed incoming mail to be opened and read by government censors before being delivered to him. Evidently Oldenburg failed to realize that his outgoing mail was also being intercepted. When he criticized the government's handling of the war with the Dutch Republic in one of his letters, he was thrown into the Tower, as a reminder not to abuse his privileges. He was released two months later, suitably chastened.

Despite all these difficulties, Oldenburg kept on publishing the *Philosophical Transactions.* It did not prove to be the financial success he had hoped for, and running it actually made him even more overworked than he had been before. But there was no doubting its influence. The *Journal des Sçavans* retreated almost immediately from its original overambitious plan to cover everything of interest to intellectuals, transforming itself instead into a scientific journal modeled on Oldenburg's. His promotion of the *Philosophical Transactions* abroad raised the international influence and authority of the Royal Society, encouraging foreign scientists to submit papers for publication and prompting traveling intellectuals to seek out the society while in London. "The fame of the Society riseth very high abroad, and makes strangers flock hither in troops; insomuch that since this March I have had no less than two dozen travellers addressed to me," Oldenburg wrote to Boyle in March 1668. By this time Oldenburg's finances were even more strained, and in 1669 the society began paying him a salary for his previously unpaid work as secretary.

As a discussion forum, a scientific journal was not as quick and immediate as a coffeehouse discussion or a society meeting. But what it lacked in speed it made up for in geographical reach. This introduced a new problem, however: how could one tell if a scientist who sent in a paper from a far-flung land was trustworthy, or whether his results were made up? Unable to query correspondents face to face, the Royal Society required that new correspondents be vouched for by existing ones who were known and trusted. In this way its network could be extended, one link at a time, through personal recommendations. In 1673, for example, the pioneering microscopist Antoni van Leeuwenhoek was introduced to the society via a letter from Reinier de Graaf, one of Oldenburg's existing Dutch contacts, who enclosed Leeuwenhoek's first written report of his observations. The letter and report were read out at one of the society's meetings and generated much interest. But its members wanted more information about this unknown amateur. A letter was duly dispatched to Christiaan Huygens's father Constantijn, asking him to visit Leeuwenhoek and assess his work and character. Huygens did so, despite the fact that England was once again at war with the Dutch Republic, and swiftly wrote back describing Leeuwenhoek as a diligent and industrious researcher. Leeuwenhoek was then invited to become a regular correspondent to the society, and he continued to send his findings for publication in the *Philosophical Transactions* until his death in 1723.

Similarly, when Willem Ten Rhijne, a Dutch doctor working in Batavia in the East Indies, wrote to the society in 1682 with some observations about the treatment of gout and the use of acupuncture among the Japanese, its members wanted to hear more. But they needed reassurance that Ten Rhijne could be trusted. Expecting this, he suggested in his letter that they consult his friend Joannes Groenevelt, a Dutch doctor living in London. Two fellows of the society duly went to meet Groenevelt, reporting back to the society the following week that Ten Rhijne had a very senior medical position in the East Indies and had been the star pupil of the chemist Franciscus (François de le Boë) Sylvius, who was well known to the society. Suitably reassured, the society added Ten Rhijne to its network of contacts and went on to publish his findings.

Oldenburg continued to publish the *Philosophical Transactions* until his death in 1677, producing 136 monthly issues in all. Two years after his death publication ceased for a while, as Hooke tried to have his own journal recognized as the official mouthpiece of the Royal Society. But his venture failed, and publication of the *Philosophical Transactions* resumed again in 1683 and has continued ever since. (The journal was formally taken over by the Royal Society in 1753.) Such was its influence that the actual meetings of the society became secondary to the conversation taking place within the pages of the journal. Unlike Newton, Hooke, and Halley, Oldenburg is little known today. He was not a scientist himself. But he made a vital contribution to the scientific revolution of the seventeenth century by creating the first modern scientific journal, a platform for international scientific collaboration that defined the template that has been used ever since.

Scientific journals distilled and formalized the discursive atmosphere of the coffeehouse, letting it reach across international boundaries via print and post to create a virtual community of scientists. Coffeehouses themselves went into decline in the late eighteenth century, as consumption of tea and coffee at home became more popular. Some transformed themselves into private members' clubs; others became mere taverns or closed down altogether.

But the vibrant, freewheeling spirit of the coffeehouse has re-emerged in recent years and now animates the Internet's discussion forums and social-media platforms. Like coffeehouses before them, they provide a free and open space for debate and discussion—perhaps too free and open, in the view of some governments, or of people who find themselves at the receiving end of online criticism or abuse. And like coffeehouses, these Internet forums initially catered to academics. When Tim Berners-Lee, a British physicist working in Switzerland, created the World Wide Web in 1991, he did so with the aim of improving communication and collaboration among his fellow researchers. As with coffeehouses, a social space first colonized by scientists turned out to have much broader appeal—and, with the endless mixing and remixing of ideas made possible by social media, has proved to be a crucible of technical and commercial innovation.

# CHAPTER 7

# THE LIBERTY OF PRINTING: THE ROLE OF SOCIAL MEDIA IN REVOLUTIONS (2)

*Without Freedom of Thought, there can be no such Thing as Wisdom; and no such Thing as publick Liberty, without Freedom of Speech.*
—Benjamin Franklin, writing as "Silence Dogood"
in the *New England Courant*, July 1722, and quoting
from the *London Journal*, 1721

## A TALE OF TWO BENJAMINS

THE INAUGURAL ISSUE of *Publick Occurrences Both Forreign and Domestick*, the first newspaper to be published in colonial North America, appeared in Boston on September 25, 1690. Its proprietor was Benjamin Harris, a local printer and coffeehouse owner who had emigrated from London, where he had got into trouble for unlicensed printing. The introduction to his paper promised "an Account of such considerable things as have arrived unto our Notice" to appear "once a month (or if any Glut of Occurrences happen, oftener)." The paper was committed to getting its facts straight. Any mistakes, it promised, would be corrected in the next issue, and anyone guilty of spreading false rumors would be named and shamed, in order to ameliorate "that Spirit of Lying, which prevails amongst us." The first issue offered a roundup of local news, including accounts of a fire, an outbreak of smallpox, and recent developments in the British colonists' war with the French to the north, in which both sides were supported by Indian tribes. Harris regarded the Indians fighting on the British side to be untrustworthy allies, denouncing them as "miserable Savages, in whom we have too much confided." This criticism, along with an offhand reference to a rumor that King Louis XIV of France had cuckolded his own son, angered the governing council of the Massachusetts Bay Colony. Its members issued a proclamation four days later to "manifest and declare their high Resentment and Disallowance of said Pamphlet, and Order that the same be Suppress'd and called in; strictly forbidding any person or persons for the future to Set forth any thing in Print without Licence." All unsold copies were destroyed, and *Publick Occurrences* never appeared again. Its first issue was also its last.

That the authorities would take a dim view of Harris's unlicensed newspaper should not have come as a surprise. Ever since the colony's founding in 1620, control of printed materials had been tight in Massachusetts, initially to ensure religious conformity. And when James II appointed Edmund Andros as the governor of New England in 1686, his instructions stated: "And forasmuch as great inconvenience may arise by the liberty of printing within our said territory under your government you are to provide by all necessary orders that no person keep any printing-press for printing, nor that any book, pamphlet or other matters whatsoever be printed without your especial leave and license first obtained." It was a similar story in other American colonies. In Pennsylvania, the governing Quakers disliked an almanac printed by William Bradford in 1685, the first item to emerge from his press, and told him "not to print any thing but what shall have Lycense from the Councill." Printing was banned outright in Virginia until the 1730s. "I thank God, there are no free-schools, nor printing; and I hope we shall not have these hundred years," the governor, William Berkeley, wrote in 1671, "for learning has brought disobedience, and heresy, and sects into the world, and printing has divulged them, and libels against the best government. God keep us from both." As a result, the colony's laws were recorded in manuscript form.

Controls on printing in the colonies did not prevent information from circulating over long distances, however. It simply traveled by letter instead. In particular, clergymen organized themselves into an informal news-sharing network that linked coastal towns with rural settlements. They were literate, well-educated, and socially connected men who knew a great deal about what was going on in their communities. In 1675–76, for example, Noah Newman, a minister in Rehoboth (part of the Plymouth Bay Colony), provided details of the conflict between Indians and colonists to other ministers. He was well placed to do so because Rehoboth was where soldiers had gathered to fight the Narrangansett and Wampanoag forces under their leader, Metacomet. Newman wrote detailed letters to John Cotton Jr., the minister in Plymouth, based on conversations with local civilians and military officers and letters they had shared with him. His reports included particulars about the time and circumstances of military action, detailed casualty lists, and eyewitness accounts from the aftermath of the fighting: "Samuel Smith's wife being big wth child & another child in her arms was crossing over an open field to a Garison house & was overtaken by the enemy & kild, & her child left alive, found standing by its dead mother where they thought it had stood neer an hour."

Cotton, who had built up an extensive network of correspondents with whom he exchanged news, copied these reports to his correspondents in Boston and Portsmouth, who copied them and passed them on in turn. One of Cotton's correspondents was his nephew Cotton Mather in Boston. In return for reports from the interior, Mather sent news from Europe back to Cotton in Plymouth, gathered from letters, conversations with merchants, and copies of the *London Gazette*. Cotton passed this information back to his network of contacts, picking specific reports and appending other news according to the interests of the recipient. In a letter to his son in 1696, for example, John Cotton Jr. copied an account from London, received from Mather, of an attempt to assassinate the king. He then added local and family news: a storm had injured some inhabitants of Deer Island, a local sailor had almost drowned after getting caught in ropes and falling overboard, and a neighbor had given birth to a daughter. Cotton had gleaned this local news from three visitors, two letters, and his own observations. Each participant in this social news network received, reported, and recopied items of interest, selecting and commenting on specific items as information traveled by letter from one person to the next.

As the informal postal service of the early colonies gave way to a more organized system, postmasters in each town emerged as important nodes in this news network, not just because they oversaw the delivery of letters, but because they were in a good position to compile and summarize information. John Campbell, who was appointed Boston's postmaster in 1702, wrote newsletters to local merchants and officials containing summaries of items culled from the London papers. As postmaster, he could send these letters without having to pay for postage. In 1704 Campbell turned his handwritten newsletter into a printed weekly, the *Boston News-Letter.* The first regular newspaper in the colonies, it was closely modeled on the *London Gazette* and soon had a circulation of around two hundred and fifty. Royal proclamations and international news appeared first, followed by news from other colonies, and finally local news. Campbell gathered information by talking to sailors, travelers, local officials, and visitors to his post office, and via handwritten newsletters from other postmasters. But most of the stories in the *Boston News-Letter* were simply copied from the London papers.

Campbell freely admitted that most of his material was not original. His aim was simply to increase the circulation of reliable information, for although "some few Gentlemen and Merchants might have all or some part thereof, yet for the most part the people in general, in this and the Neighbouring Provinces, have it not, and what they have, variously, and often falsely reported." So committed was Campbell to providing a full account of international news, in fact, that any items that would not fit were held until the next issue. The result was a growing backlog of stories, and eventually the foreign reports in the *Boston News-Letter* were almost a year out of date. Campbell saw his role as providing an unbroken thread of historical reports, rather than the latest information. He was not interested in expressing his own opinions or reporting on local politics, and he had the contents of every issue approved in advance by the governor's office to ensure that he would not get into trouble.

The *Boston News-Letter* was delivered directly to its subscribers in town and sent by mail to those farther afield. Items from its pages, and sometimes whole copies, were passed in turn along existing letter-exchange networks. Samuel Sewall, a Boston minister who regularly sent a digest of news to his network of other ministers across New England, began to include copies of the *Boston News-Letter* with his letters. Sometimes he would send three or four copies of a single issue so that the recipients of his letters could keep one and mail the others on to their own correspondents. The advent of the *Boston News-Letter* meant that Sewall could focus his letters on local news, family news, and his own comment and analysis, because the basic facts were covered in print. He and other ministers wrote about the implications of news stories for themselves as well as for their churches and communities. John Campbell even gave his subscribers the option to receive the *Boston News-Letter* printed on a full sheet of paper, rather than the customary half sheet, so that personal commentary could be added by hand on the blank half sheet. In these early days of small-circulation local papers, newspapers and letters were complementary forms of communication, each depending on the other, and mixing information from letters, conversations, and other printed sources. The resulting circulation of printed and written information enabled the community to talk to itself.

The *Boston News-Letter* inspired a rival paper in Boston, the *Boston Gazette*, launched in 1719. It also spawned imitators elsewhere, including the *American Weekly Mercury* in Philadelphia and the *New-York Gazette*. What all these papers had in common was an unwillingness to offend the authorities, which manifested itself in a bland prose style and an avoidance of local political news. The governor of Massachusetts from 1702 to 1715, Joseph Dudley, was deeply unpopular, but you would never have known it from the pages of the Boston papers.

Gradually, though, the American press started to become more open and adversarial. A handful of pamphlets printed between 1700 and 1706, written by Samuel Sewall, Cotton Mather, and others, argued about whether Christianity was compatible with slavery. Then in 1707, as part of a long power struggle between the Mather family and Governor Dudley, the Mathers had two pamphlets published in London describing Dudley's mismanagement of the colony and accusing him of illegally trading with the French. Copies of these pamphlets began to circulate in New England, and one of them, "A Memorial of the Present Deplorable State of New England," was reprinted in Boston. But Dudley responded with a pamphlet of his own, "A Modest Enquiry into the Grounds and Occasions of a Late Pamphlet," in which he rebutted the charges with entirely uncharacteristic confidence and grace, scuppering the Mathers' attempt to have him recalled to London.

In 1720 a pamphlet fight erupted between two Bostonian politicians, Elisha Cooke Jr. and Joseph Dudley's son Paul, over the extent to which the governor, appointed by the king, could encroach upon colonists' rights. Cooke, who had fallen out with Joseph Dudley's successor as governor, Samuel Shute, fired the opening shot with a pamphlet objecting to the governor's power of veto over the decisions of the Massachusetts legislature. Drawing inspiration from the satirical works of Daniel Defoe, Paul Dudley responded with a pamphlet set on the mythical island of "Insania," where a noble governor finds himself under attack from foolish "Country Men" who wish to sow confusion and discord. Cooke responded with a pamphlet in the form of a dialogue between a "Country Gentleman" and a "Boston Gentleman," who argue about whether true patriotism should be defined as obedience to the king or love of one's country. Other pamphleteers also weighed in on each side, and the resulting furor prompted Governor Shute to return to England in 1723 to demand clarification of the extent of his powers. This exchange of pamphlets—"almost an infinite number of Pamphlets dispersed thro' the Country," as Cooke called it in one of his works—was an early manifestation of colonial opposition to overreach from London.

An outbreak of smallpox in Boston in 1721 prompted another war of words, as people argued about the merits of inoculation, then a controversial new treatment. Cotton Mather had learned about the practice from his slaves and wrote an essay, which was circulated in manuscript form among local doctors, advocating its use. (Inoculation at the time involved deliberate infection of a small cut in the arm of the patient, to increase immunity to smallpox.) A local doctor, Zabdiel Boylston, started performing inoculations, and his endorsement by Mather and several Puritan clergymen was seen by other Boston doctors, who were skeptical about the merits of inoculation, as an incursion onto their professional turf. Letters arguing for and against inoculation appeared in the *Boston News-Letter* and *Boston Gazette*, along with a series of pamphlets, written under a multitude of pseudonyms, supporting each side of the argument. In August 1721 the doctor at the head of the anti-inoculation faction, William Douglass, went so far as to set up a new Boston newspaper, the *New-England Courant*, to champion his side of the argument and criticize his Puritan opponents.

Unlike the *Boston News-Letter* and the *Boston Gazette*, this new paper was not approved by the authorities before publication. Rather than being a tedious gazette, it drew inspiration from the more argumentative and opinionated *Spectator* in London. Under the editorship of James Franklin, who was assisted by his apprentice and younger brother Benjamin, the *New-England Courant* issued a relentless stream of anti-inoculation propaganda. Formal editorializing was rare in early American papers, and opinions were instead presented by publishing letters to the editor—though, in some cases, the editors wrote the letters themselves under pseudonyms. In this way a local paper acted as a forum for the opinions of both its readers and its editors. As James Franklin put it in the *New-England Courant*, "I hereby invite all Men, who have Leisure, Inclination and Ability, to speak their Minds with Freedom, Sense and Moderation, and their Pieces shall be welcome to a Place in my Paper." In practice, of course, he chose what to print. And even then the custom of writing under pen names could spring surprises. Benjamin Franklin's first foray into newspaper writing took the form of a series of fourteen letters, supposedly written by a widow named Silence Dogood, which he submitted anonymously to the *New-England Courant*'s office, and which were enthusiastically published by his unwitting brother. James was furious when sixteen-year-old Benjamin admitted to having written the letters. This tale does not simply illustrate Benjamin's ingenuity and writing prowess; it also shows how newspapers at the time were open to submissions from anyone, provided they expressed an interesting opinion. Such newspapers consisted in large part of reprinted letters, speeches, and pamphlets, and thus provided a platform by which people could share and discuss their views with others. Early newspapers were, in short, a form of social media.

The argument over inoculation raged for the rest of 1721 in Boston's newspapers and pamphlets. The two sides marshaled religious, legal, and medical arguments, drew up affidavits from patients, and made personal attacks on their opponents. At one point Mather denounced the *New-England Courant* as a "Notorious, Scandalous paper . . . full-freighted with Nonsence, Unmannerliness, Railery, Prophaneness, Immorality, Arrogancy, Calumnies, Lyes, Contradictions, and what not, all tending to Quarrels and Divisions, and to Debauch and Corrupt the Minds and Manners of New England." We now know that Mather and the supporters of inoculation were right, and the opposing doctors were wrong. Of the 10,500 people living in Boston in 1721, around 6,000 contracted smallpox and 900 died, an overall mortality rate of 15 percent. The mortality rate among the 286 people who were inoculated, by contrast, was just 2 percent.

The authorities stayed out of that fight, but James Franklin got into trouble when he published a seemingly harmless paragraph the following year: "We are advised from Boston that the government of Massachusetts are fitting out a ship, to go after the pirates, to be commanded by Captain Peter Papillon, and 'tis thought he will sail some time this month, wind and weather permitting." The implication of this sarcastic report was that the authorities were not fully committed to combating piracy along the New England coast, and James was jailed for three weeks until he apologized. During this time Benjamin ran the *New-England Courant* in his brother's place.

The following year James fell foul of the authorities again, this time for running a letter insulting the clergy. He was told that he could only continue to print his newspaper if he submitted each edition for approval beforehand, as the other Boston papers did. But James spotted a loophole in the ruling: it did not prevent other people from printing the newspaper without approval, so he named Benjamin the publisher instead and went into hiding. Benjamin later recalled that "during my Brother's Confinement, I had the Management of the Paper, and I made bold to give our Rulers some Rubs." This farcical situation continued until late 1723, at which point the Franklin brothers parted ways. The *New-England Courant* folded two years later. It had been wrong about inoculation, but it had pioneered a combative, campaigning use of the press and a more entertaining, literary style. Most importantly, it had inspired similar newspapers that also refused to submit to prepublication approval and dared to be critical of the colonial authorities, such as the *New-England Weekly Gazette* in Boston and the *Pennsylvania Gazette*, launched by Benjamin Franklin in Philadelphia. By 1730 all attempts to enforce licensing had been abandoned, and the orders granted to colonial governors no longer made any mention of it.

Printers, pamphleteers, and writers could still be prosecuted after publication if their output displeased the authorities, however, as the case of John Peter Zenger showed. He was a printer who was paid to launch a new paper, the *New-York Weekly Journal*, in 1733 in order to expose the abuses of power by New York's new governor, William Cosby. Among other things, Cosby rigged elections, awarded generous pensions to his cronies, and claimed that he was entitled to half the pay awarded to the official who had governed on his behalf before his arrival from Europe. The chief justice, who stood up to him, was fired. Cosby felt that holding a position of authority gave him the right to do as he pleased: in a previous job as governor of the island of Minorca, he had illegally seized a Portuguese ship and tried to claim its cargo for himself. Zenger's paper published a series of increasingly vitriolic letters criticizing Cosby's behavior, accusing him of threatening the "liberties and properties" of the people and of violating the rules of his office. Unable to identify the author of these attacks on his person, Cosby ordered Zenger's arrest instead. After refusing to identify the author of the offending letters, Zenger spent eight months in jail and was then put on trial for seditious libel.

Zenger was defended by Andrew Hamilton, one of the finest lawyers of his generation, and a friend of Benjamin Franklin, who probably asked him to do it. Hamilton's defense strategy was highly unusual. Under English law, the truth of a statement was not a defense against libel, so it was irrelevant whether or not the accusations made against Cosby were true; all that mattered was whether Zenger had, indeed, printed them. Hamilton immediately admitted that Zenger had, seemingly conceding the case. But he went on to argue that it was the law itself that was at fault, and asked the members of the jury to assess for themselves whether the accusations against Cosby were accurate, arguing that "the words themselves must be libelous, that is, false, scandalous, and seditious, or else we are not guilty." To general astonishment, the jury agreed with him, acquitting Zenger, infuriating Cosby, and causing a sensation throughout the colonies. By showing that juries were unwilling to convict in cases of seditious libel, the case removed the last means available to colonial governors to control the press. Writers and publishers finally had a free hand to print what they wanted.

Benjamin Franklin had helped to bring this about by pioneering unlicensed printing at the *New-England Courant* and, probably, by asking Hamilton to defend Zenger. He went on to make a further contribution to the free flow of information in the colonies. In 1737 he was offered the job of Philadelphia's postmaster, replacing the editor of a rival newspaper in the town who had run the postal service poorly and refused to allow postal riders to deliver Franklin's paper. "I accepted it readily," Franklin later wrote in his autobiography, "and found it of great advantage; for, tho' the salary was small, it facilitated the correspondence that improved my newspaper." He ran the system so efficiently that in 1753 he was made deputy postmaster general for North America. In this role he increased the reliability and frequency of the postal service, reorganizing routes and streamlining procedures. The number of deliveries from New York to Philadelphia went from one a week to three a week, and the postal service became profitable for the first time. Just as important, Franklin allowed free exchange by post of newspapers both within and between colonies, formalizing the reprinting of noteworthy reports and letters by papers in different towns. As the publisher and printer of the *Pennsylvania Gazette*, encouraging the circulation of news in these various ways was in Franklin's own commercial interests. But it also contributed to the dynamism, vitality, and unity of the American colonies' emerging information ecosystem, with its constant traffic of letters, pamphlets, and newspapers. By the 1760s it had developed into a powerful, open, and social platform for the rehearsal of arguments, the propagation of ideas, and the exchange of opinions. It was both powerful and highly valued, as became apparent when relations with Britain started to break down and the media system itself was threatened by taxation.

## "THE ART OF PRINTING SHOULD BE ENCOURAGED"

In May 1765 news reached the American colonies that the government in London had passed "an act for granting and applying certain Stamp Duties, and other Duties, in the British Colonies and Plantations in America, towards further defraying the Expenses of defending, protecting, and securing the same." The British government had borrowed heavily to fund a war with the French around the world, which had been fought in Europe, the Americas, the Philippines, and parts of Africa and India. In Europe this conflict was known as the Seven Years' War; in America, where French and English colonists fought each other with the backing of different Indian tribes, it was called the French and Indian War; and in retrospect it was arguably the first true world war. The government in London reasoned that having fought, in part, to protect its American colonists, it was only right that those colonists should pay some of the cost of defending their territory. The Stamp Act was intended to raise about one hundred thousand pounds a year, or a bit less than half the expected cost of maintaining a garrison in North America. Starting in November 1765, government-appointed agents would sell specially stamped paper. Legal documents that did not appear on paper marked with an appropriate stamp would not be considered binding by the courts, which would encourage compliance with the law, and it would be illegal to print newspapers or pamphlets on unstamped paper. From the point of view of the government in London, the merit of a stamp tax was that it would be relatively simple to administer and enforce.

The colonists saw things rather differently. The full text of the act was quickly reprinted in Philadelphia, Boston, Annapolis, New London, New York, and New Jersey. It could not have been better designed to antagonize lawyers, merchants, and printers, who were some of the most influential and outspoken members of society. Wills, property deeds, contracts, insurance policies, and other legal documents were subject to varying amounts of stamp duty; the tax was two pounds on a degree certificate, four pounds on a wine-seller's permit, and ten pounds on a lawyer's license. Printers would be required to pay a penny in tax for each sheet of paper used to print newspapers, and two shillings for every advertisement carried, roughly equivalent to a fifty-percent tax on advertising revenue. A similar tax was imposed on paper for printing pamphlets, increasing with the number of pages. Playing cards were to be taxed at a shilling a pack, and dice ten shillings. These taxes had to be paid with hard currency, rather than colonial paper currency.

All this would be inconvenient and costly, but it was also a direct attack on the colonists' ability to exchange information with each other. Worst of all, it would force the colonists to pay taxes levied by a parliament in which they were not represented. A previous effort to tax the colonists by imposing duty on molasses imported from French islands in the Caribbean had been ignored, but the government had started to enforce its collection under the Sugar Act, passed in 1764. This was unpopular, not least because it indirectly increased the price of rum, and led to protests in Boston. But the Stamp Act went much farther. Its imposition of direct and highly visible taxes on a wide range of everyday items and activities provoked opposition throughout the colonies. At a demonstration against the stamp tax in Boston, the man appointed as the local stamp distributor was burned in effigy, prompting him to resign from the post. Accounts of the protest rippled through the newspapers to other colonies and triggered similar actions elsewhere.

Opponents of the tax took up their pens to write pamphlets and letters to newspapers. Benjamin Church, a Boston doctor and poet, wrote a pamphlet entitled "Liberty and Property Vindicated, and the Stampman Burnt" in which he declared "my opinion is that it cannot be counted rebelling for the freemen of this colony to stand for their absolute rights and defend them (as a man would his own house when insulted)." A pamphlet published by a lawyer in North Carolina,

Maurice Moore, led to the resignation of the stamp distributor and the abandonment of local attempts to enforce the act. In Annapolis a lawyer named Daniel Dulany published a pamphlet entitled "Considerations on the Propriety of Imposing Taxes in the British Colonies" that met with widespread acclaim for its arguments against taxation without representation. Dulany regarded the Stamp Act as a mistake by the British government, but he was opposed to the notion of independence, which some colonists had begun to advocate openly. His pamphlet prompted an exchange of open letters with Charles Carroll, a wealthy Maryland planter who favored independence and insisted on the right of the colonies to determine their own taxation. These letters, written under the pseudonyms "Antillon" and "First Citizen" respectively, appeared in the *Maryland Gazette* and became increasingly vociferous. Carroll had the better of the argument and, after his identity was revealed, went on to become a leading figure in the campaign for independence.

The effectiveness of newspapers and pamphlets in stirring up opposition to the Stamp Act led to calls for those attacking the policy, such as the publishers of the newly radicalized *Boston Gazette*, to be prosecuted for seditious libel. In a series of anonymous letters published in the *Boston Gazette*, John Adams, a Boston lawyer, responded that this was just one more example of how the Stamp Act was infringing upon the colonists' rights, by interfering with the free flow of information itself:

> Care has been taken that the art of printing should be encouraged, and that it should be easy and cheap and safe for any person to communicate his thoughts to the public. And you, Messieurs printers, whatever the tyrants of the earth may say of your paper, have done important service to your country by your readiness and freedom in publishing the speculations of the curious . . . the jaws of power are always opened to devour, and her arm is always stretched out, if possible, to destroy the freedom of thinking, speaking, and writing . . . Be not intimidated, therefore, by any terrors, from publishing with the utmost freedom, whatever can be warranted by the laws of your country . . . it seems very manifest from the Stamp Act itself, that a design is formed to strip us in a great measure of the means of knowledge, by loading the press, the colleges, and even an almanack and a newspaper, with restraints and duties.

An unprecedented sense of cooperation and collective resistance emerged among the colonists, culminating in the Stamp Act Congress in New York in October 1765. Delegates from nine colonies met to call for the act's repeal, and to insist that only colonial assemblies had the right to levy taxes. The meeting's final declaration was drawn up by John Dickinson, a lawyer representing Pennsylvania, who also wrote two pamphlets arguing against the act. One of them, "The Late Regulations respecting the British Colonies on the Continent of America Considered, in a Letter from a Gentleman in Philadelphia to his Friend in London," was aimed at British readers and caused sufficient interest that it was reprinted in London. The other pamphlet, reprinted in the *New-York Gazette* and other newspapers, urged Pennsylvanians to ignore the new law, explaining that it was a test of their willingness to stand up to tyranny. "Your Conduct at this Period must decide the future Fortunes of yourselves and of your Posterity—must decide whether Pennsylvanians from henceforward shall be Freemen or Slaves," he wrote. "If you comply with the Act by using Stamped Papers, you fix, you rivet perpetual Chains upon your unhappy Country."

The argument was not entirely one-sided. Loyalist views also received an airing: writing under the name "Americanus" a Pennsylvania politician, Joseph Galloway, argued that it was only right that the colonists should contribute to their own defense, and that they were unwise to pick a fight with the British government. "Permit me, however unpopular the task, through the impartial channel of your paper, to point out the impudence and folly of such conduct," he wrote. His letter appeared in the *New-York Gazette* and the *Pennsylvania Journal*. But for the most part the presses groaned with pamphlets and newspapers denouncing the Stamp Act—not least because printers themselves were strongly opposed to it.

In the weeks before the act came into force, the *Maryland Gazette* appeared with a skull-and-crossbones on its front page, and restyled itself "The Maryland Gazette Expiring." Many newspapers ceased publication on November 1, when the act took effect. But they gradually reemerged as it became clear that opposition to the law was so widespread that it could not be enforced. The *Pennsylvania Gazette* reappeared on November 7, on unstamped paper, with the words "No Stamped Paper to be had" in place of its masthead. Attempts to enforce the law were abandoned and the British government repealed it in March 1766. The news was greeted with jubilation in America and led to a further outpouring of pamphlets, sermons, and celebratory poems. But on the same day it repealed the Stamp Act, the British Parliament passed another act insisting on its authority to make laws binding the colonies. This led in 1767–8 to the Townshend Acts which, among other things, imposed taxes on goods imported to America from England, tightened up the enforcement of customs duties, and appointed new tax collectors and inspectors. These measures were passed in five separate acts in 1767 and 1768 and their effects were more widely dispersed than that of the Stamp Act, so at first they did not generate the same level of opposition.

This inspired John Dickinson to take up his pen once again in late 1767, this time to write a series of twelve anonymous letters in the persona of a farmer, which first appeared in the *Pennsylvania Chronicle.* Such was their popularity that they were reprinted in nearly all the newspapers throughout the colonies and then published as a pamphlet, "Letters from a Farmer in Pennsylvania to the Inhabitants of the British Colonies," which was originally printed in Philadelphia and then reprinted in Boston, New York, and Williamsburg, Virginia. Dickinson warned his readers of the dangers posed by the new laws, calling upon them to "rouse yourselves, and behold the ruin hanging over your heads." By requiring the colonies to import certain goods (including glass, paper, and tea) from Britain and prohibiting their manufacture in America, the government was taking away their liberty: "If Great Britain can order us to come to her for necessaries we want, and can order us to pay what taxes she pleases before we take them away, or when we land them here, we are as abject slaves," he wrote. In subsequent letters he explained the consequences of the other Townshend Acts: paying officials' salaries directly from London, for example, undermined the authority of the local governing councils.

Dickinson's letters laid out the political arguments in a digestible and easily understandable form, though he stated clearly that he disapproved of resistance by force. His writings helped marshal opposition to the Townshend Acts, in the form of coordinated nonimportation agreements adopted across the colonies. The boycott of British goods was enforced via the press, and merchants who violated it were named and shamed in newspapers and in broadsides posted in public places. The boycott, along with the uproar that followed the killing of five protesters by British soldiers in Boston in 1770, led the government to repeal the Townshend Acts. Only the duty on tea was left in place as a symbol of the government's insistence on its right to tax the colonists. But it was easily avoided by buying tea smuggled in from the Netherlands instead.

The uneasy calm that followed was broken in 1773 with the passage of the Tea Act, an attempt to impose taxation on the American colonists by stealth. It granted the East India Company a monopoly on tea-selling and hid an import duty of three pence per pound in the price, which the East India Company would pay directly to the government. It did not take long for the colonists to work out what was going on. Broadsides appeared announcing public meetings, and resolutions condemning the act were passed and then published in the newspapers, prompting similar actions in other towns. As with the opposition to the Stamp Tax, a coordinated campaign forced the designated tea-sellers in each colony to resign, and to prevent the East India Company's ships from unloading their tea. In Boston, however, the governor refused to turn the tea ships away, and the tea-sellers, two of whom were his sons, would not resign. The result was the Boston Tea Party, planned in the house of Benjamin Edes, the proprietor of the *Boston Gazette*, which had by this time become the city's most radical newspaper. On the evening of December 16 a group of men, some of whom stopped off in the *Boston Gazette*'s offices to don disguises as Mohawk Indians, boarded the three tea ships and tipped all 342 chests of tea on board into the water.

The Boston Tea Party was celebrated in songs, poems, engravings, and pamphlets. The British government responded with the so-called Coercive Acts—a series of punitive laws intended to assert its authority, which served only to enrage the colonists further. The colonists responded by calling the First Continental Congress to coordinate their actions. Armed conflict now seemed inevitable. In a heated newspaper debate between "Novanglus" (John Adams, writing on behalf of disgruntled colonists) and "Massachusettensis" (Daniel Leonard, defending the British government's position), the tone of the argument became steadily more confrontational. Over the course of twelve letters published in the Boston newspapers between January and April 1775, Adams outlined what he regarded as a conspiracy between the British government and its aristocratic supporters in America to deprive the colonists of their rights and liberties. Adams idolized

Cicero, and his letters were inspired by the Roman statesman's denunciation of the Catilinarian conspiracy of 63 B.C., one of the most famous speeches in classical literature. Adams insisted that the colonists simply wanted to protect their rights and maintain their autonomy. "The patriots of this province desire nothing new; they wish only to keep their old privileges," he wrote. "They were for one hundred and fifty years allowed to tax themselves and govern their internal concerns as they thought best." The last letter Adams wrote as Novanglus never appeared, however, because hostilities began at Lexington and Concord on April 19. As Adams later put it, this "changed the Instruments of Warfare from the Pen to the Sword." Yet even after the Revolutionary War had begun, the pen would still have a role to play.

## HOW THOMAS PAINE WENT VIRAL

The most popular and influential publication of the American colonial media system appeared in January 1776 from an unlikely author. Thomas Paine had emigrated from England a little more than a year earlier. Aged thirty-seven, he had worked in various jobs as a corset maker, teacher, and excise officer. After losing his job, separating from his wife, and having to sell his possessions to avoid debtor's prison, he was introduced to Benjamin Franklin, then in London, who suggested that he emigrate to America. Paine arrived in Philadelphia at the end of November 1774 and soon got a job as the editor of the *Pennsylvania Magazine*, where his newly discovered talent as a provocative writer and editor provided the ideal outlet for his strongly held Enlightenment views. Paine wrote under various pseudonyms, including "Humanus," "Atlanticus," and "Vox Populi," initially to conceal the small magazine's number of writers, but then as a means of expressing increasingly radical positions. When his articles were deemed too extreme by the magazine's proprietor and printer, Paine would submit them to the *Pennsylvania Journal* instead, as he did with his searing attack on slavery, published in March 1775. It led to the establishment of the first antislavery society in America the following month.

By the autumn of 1775, however, Paine had fallen out with his proprietor and began work instead on a long essay, which he planned to publish as a series of letters in the newspapers, about the colonies and their relationship with Britain. But what was originally intended as a historical account came out instead as a forthright and detailed argument in favor of American independence. Paine decided to publish it as a pamphlet, under the title "Plain Truth," but he was persuaded by a friend to rename it *Common Sense*. Because of its incendiary subject matter only one printer in Philadelphia, Robert Bell, would print it, and he demanded unusual terms. Paine would be liable for any financial loss if the pamphlet failed to sell, but would share any profits with Bell if it did well. Bell also set the cover price at a steep two shillings—higher than the going rate for a pamphlet of its size. Despite this impediment, the first thousand copies sold out within days of the pamphlet's publication on January 10, 1776. Having fallen out with Bell, who continued to print the pamphlet anyway, Paine took his pamphlet to another printer, ordering six thousand copies to be printed and sold at the more reasonable price of one shilling, with any proceeds to be donated to the colonists' Continental Army, led by George Washington. This gesture, along with the intense interest the pamphlet had already generated, its low cost, and newspaper advertisements taken out by the two rival printers, each attacking the other, only served to heighten interest in it.

As a latecomer to the colonies' struggle with Britain and steeped in the philosophy of the Enlightenment, Paine saw the situation with stark clarity: it seemed obvious to him that the colonies should cut their ties with the king and his government, abandon the European traditions of hereditary monarchy and class-based societies, and strike out on their own. His own experience led him to reject the traditional patriotic view that the British constitution and monarchy was the greatest form of government yet devised, and he argued that trying to find some conciliatory compromise with Britain would lead only to the "ruin of the continent." Far better, he argued, for the colonists to declare independence and devise a new form of government, based on both state-level and national assemblies. These were hardly new ideas, but Paine articulated them more clearly and forcefully than anyone had done before. He adopted a far more plain-spoken, populist, and angry tone than that usually taken in pamphlets, in which opponents would often not be mentioned by name. Britain, he said, was "now an open enemy." He called George III the "Royal Brute of Britain" and the "sullen-tempered Pharaoh of England." He did not assume any learning or familiarity with the classics, or with contemporary political philosophy, in his readers, but modulated his voice and simplified his language, as Luther had done, to reach the widest possible audience. Paine presented the choice that the colonists faced as a historical turning point: they had an opportunity, he argued, to make a complete break with the past and "begin the world anew." He distilled the argument for independence into a form that had widespread appeal: the colonists, on the side of good and the future, versus the British, on the side of evil and the past.

*Common Sense* appeared just as the Second Continental Congress, the body established to coordinate the actions of the thirteen colonies, was inching toward embracing the idea of independence. But its delegates were uncertain whether they had the authority to make such a decision and were unsure of the extent of popular support for the idea. Even after the outbreak of hostilities at Lexington and Concord, many delegates still favored reconciliation with Britain, coupled with an agreement to protect the colonists' rights. Independence was regarded as a radical and dangerous policy. Washington himself had said in May 1775 that "one might set me down for everything wicked" if he came out in favor of independence. Supporters of independence, meanwhile, such as John Adams, one of the delegates for Massachusetts, worried that Paine's outspoken attack on the monarchy would discredit their position. Paine's pamphlet was, in short, regarded as dangerously extremist, and early readers who were convinced by its arguments were sometimes unsure whether they ought to express their enthusiasm for it.

Within ten days of its appearance, *Common Sense* had spread to New York, Virginia, and Massachusetts. Starting with the delegates to the Second Continental Congress in Philadelphia, copies rippled through the political class, passing from hand to hand, sent and recommended to friends in the post. Samuel Adams, another Massachusetts delegate (and cousin of John Adams), sent a copy of the pamphlet to his wife in Boston by letter, and wrote to his friend James Warren suggesting that he borrow it from her: "I have sent to Mrs. Adams a Pamphlet which made its first Appearance a few days ago. It has fretted some folks here more than a little. I recommend it to your Perusal and wish you would borrow it of her." Warren, it turned out, had already received several copies from others in Philadelphia, which he immediately shared with his family, neighbors, and friends, one of whom wrote that he was "much obliged for the loan of it" and advocated that

COMMON SENSE;

ADDRESSED TO THE

INHABITANTS

OF

AMERICA,

On the following intereſting

SUBJECTS.

I. Of the Origin and Deſign of Government in general, with conciſe Remarks on the Engliſh Conſtitution.

II. Of Monarchy and Hereditary Succeſſion.

III. Thoughts on the preſent State of American Affairs.

IV. Of the preſent Ability of America, with ſome miſcellaneous Reflections.

---

Man knows no Maſter ſave creating HEAVEN,
Or thoſe whom choice and common good ordain.

THOMSON.

---

The cover of Thomas Paine's pamphlet *Common Sense*, which was initially published anonymously. *Library of Congress* its arguments be "republished in all the newspapers." Josiah Bartlett, a delegate from New Hampshire, sent a copy to a friend back home explaining that it was being "greedily bought up and read by all ranks of people" and "which you will please to lend round to the people." A delegate from Virginia sent a copy to Thomas Jefferson, with the words "I send you a present of 2 shillings worth of Common Sense." Some delegates wanted to see what reaction the pamphlet would provoke before revealing their own opinions. But others were happy to admit that they had been persuaded by Paine's arguments right away. His pamphlet both revealed the extent of latent support for independence and won new converts to the cause.

In late January 1776, General Charles Lee wrote to Washington: "Have you seen the pamphlet 'Common Sense'? I never saw such a masterly irresistible performance. It will, if I mistake not, in concurrence with the transcendent folly and wickedness of the ministry, give the coup-de-grace to Great Britain. In short, I own myself convinced, by the arguments, of the necessity of separation." Washington wrote to a friend that "the sound doctrine and unanswerable reasoning contained in the pamphlet 'Common Sense' will not leave members [of the Second Continental Congress] at a loss to decide upon the propriety of separation." He wrote to a friend in April that "by private letters which I have lately received from Virginia, I find 'Common Sense' is working a wonderful change there in the minds of many men."

In Philadelphia, Paine increased the circulation of the pamphlet by allowing more printers to copy it. He was glad to see it being reprinted elsewhere, as booksellers and printers rushed to capitalize on its popularity. It had soon been reprinted in New York; Salem, Massachusetts; Hartford, Connecticut; Lancaster, Pennsylvania; Albany, New York; Providence, Rhode Island; and Norwich, Connecticut. Copies of the reprint in Salem were sold "by the bundled dozen." *Common Sense* was now being read aloud in homes, shops, taverns, and coffeehouses. Its heavy use of italics and commas helpfully indicated where stress should be placed and breath should be taken. It was even read out as the Sunday sermon in a Connecticut church. By the end of March it had sold one hundred thousand copies, and it eventually sold around two hundred and fifty thousand, making Paine the world's best-selling author. Extracts of *Common Sense* also appeared in newspapers throughout the colonies. The *Norwich Packet*, a New

England paper, excerpted it over nine consecutive issues between February and April. The *Connecticut Courant* reprinted all but its first section over three issues, and two Virginia papers also ran extracts.

The success of *Common Sense* transformed attitudes toward independence, which had previously been something many people were reluctant to talk about, let alone support. A Bostonian reader of the pamphlet wrote to a friend that "independence a year ago could not have been publickly mentioned with impunity . . . Nothing else is now talked of, and I know not what can be done by Great Britain to prevent it." A letter published in the *Boston Gazette* on April 29 declared that "had the spirit of prophecy directed the birth of a publication, it could not have fallen upon a more fortunate period than the time in which 'Common Sense' made its appearance. The minds of men are now swallowed up in attention to an object the most momentous and important that ever yet employed the deliberations of a people." Another reader, in Maryland, whose letter was printed in the *Pennsylvania Evening Post*, declared that the author of *Common Sense* (who was still, at this point, anonymous) "has made a great number of converts here. His style is plain and nervous; his facts are true; his reasoning just and conclusive." An open letter to the author of *Common Sense* in the *New London Gazette* was just as complimentary: "In declaring your own, you have declared the sentiments of Millions. Your production may justly be compared to a landflood that sweeps all before it. We were blind, but on reading these enlightening works the scales have fallen from our eyes." The *Connecticut Courant* reported in April that "May the INDEPENDENT principles of COMMON SENSE be confirmed throughout the United Colonies" had become a favorite toast "in the best companies."

Not everyone agreed with Paine, of course. A rival pamphlet called "Plain Truth," authored by "Candidus," used quotations from a range of political philosophers to attack Paine's arguments. This may have impressed the aristocratic loyalists for whom it was written, but like the Latin rebuttals to Luther's works, it failed to connect with a wider audience. Indeed, it attracted a response of its own, "Remarks on a Late Pamphlet entitled Plain Truth," which, while expressing sympathy with its argument, suggested that it was so badly written that it would only strengthen the cause of independence. "Plain Truth" was such a flop that Paine himself remarked that it "hath withered away like a sickly unnoticed weed, and which, even its advocates are displeased at." Just as unsuccessful was "The Deceiver Unmasked," written by the clergyman of a loyalist congregation in New York. When its availability was advertised in the New York papers, supporters of independence descended upon the printer's workshop and destroyed all remaining copies of it.

*Common Sense* inspired heated commentary in the newspapers, from both supporters and detractors. The author of a letter to the *Pennsylvania Packet* remarked upon "the manner of conducting what is now called the Independent controversy in the newspapers." The most effective response was that of William Smith, a loyalist clergyman who wrote a rebuttal to *Common Sense* in the form of a series of eight letters to the *Pennsylvania Gazette* under the name "Cato." Smith professed himself open to independence at some vague point in the future, but he believed reconciliation and compromise with the British was still the best policy. The fourth letter in the series provided the most detailed and penetrating response to Paine's call for independence, and Paine's friends, including Benjamin Franklin, convinced him that it could not go unanswered. Paine responded with four letters of his own, published in the Pennsylvania newspapers, signed "The Forester." Both sets of letters were widely reprinted in other papers and prompted responses from others. (Some people even complained that there was no room for anything else in the Philadelphia newspapers.) Paine ridiculed Cato's writing as "insipid in its style, language, and substance" and then criticized the weakness of his arguments. Neither writer emerged as the clear victor in print. But as it became clear that public sentiment was shifting in favor of independence, Smith realized he had lost the argument.

What had seemed unthinkable in July 1775, when the colonists had sent a petition to London seeking reconciliation, had come to appear inevitable a year later, and on July 4, 1776, the Second Continental Congress adopted the Declaration of Independence. In the following months and years Paine wrote a series of pamphlets entitled "The American Crisis" to boost morale during the Revolutionary War and argue against any compromise with the British. The first appeared in the dark days of December 1776, after Washington's Continental Army had suffered a series of reversals. It contained the stirring passage: "These are the times that try men's souls. The summer soldier and the sunshine patriot will, in this crisis, shrink from the service of their country; but he that stands by it now, deserves the love and thanks of man and woman. Tyranny, like hell, is not easily conquered; yet we have this consolation with us, that the harder the conflict, the more glorious the triumph." Washington ordered his officers to read Paine's new pamphlet to his men to inspire them, and two days later he led his men across the Delaware River to win victories at Trenton and Princeton. Paine went on to produce subsequent editions of "The American Crisis," responding to each twist and turn in the conflict that eventually led to Britain's recognition of the United States of America in 1783.

But *Common Sense* was his masterpiece. It was unquestionably the most popular and influential pamphlet of the American Revolution. "In unison with the feelings and sentiments of the people, it produced astonishing effects," wrote David Ramsay, a doctor from South Carolina who published one of the earliest accounts of the American Revolution. "It was read by almost every American, and in conjunction with the cruel policy of Great Britain, was by the direction of Providence, instrumental in effecting an unexampled unanimity in favor of independence." Like Luther's pamphlets in the sixteenth century and Internet-based social media in the twenty-first, *Common Sense* acted as a means of signaling and synchronizing opinion. Its arguments, and the discussions they sparked, helped reveal the widespread support for independence among the colonists, both to each other and to their political leaders.

Many years later John Adams wrote disapprovingly to Thomas Jefferson that "history is to ascribe the American Revolution to Thomas Paine." That is an exaggeration, but not much of one. The revolution was certainly helped along by America's unusually free and open social-media ecosystem. The circulation of letters, pamphlets, and newspapers and the resulting interchange of ideas enabled the separate colonies to unite behind a common cause. As William Cushing, a Massachusetts judge, asked in a letter to John Adams in 1789: "Without this liberty of the press could we have supported our liberties against British administration? Or could the revolution have taken place? Pretty certainly it could not, at the time it did." In *The History of the American Revolution*, published in 1789, Ramsay came to the same conclusion. "In establishing American independence," he declared, "the pen and the press had merit equal to that of the sword."

# CHAPTER 8

# THE SENTINEL OF THE PEOPLE: TYRANNY, OPTIMISM, AND SOCIAL MEDIA

*Where does so much mad agitation come from? From a crowd of minor clerks and lawyers, from unknown writers, starving scribblers, who go about rabble-rousing in clubs and cafés. These are the hotbeds that have forged the weapons with which the masses are armed today.*

—P. J. B. Gerbier, June 1789

## READING THE FRENCH MEDIA

IN 1630 THÉOPHRASTE Renaudot, a French doctor, opened an unusual office in Paris. The Bureau d'Adresse et de Rencontre was intended to act as a clearinghouse for information. People looking for work could go to the office and sign up in its register, for example, and anyone looking to hire someone could then examine its pages to look for suitable employees. Registering with the office involved paying a small fee, which was waived for the very poor. The Bureau was given a royal patent granting it a monopoly, and by 1639 a decree had made it compulsory for all workmen entering Paris to register at the Bureau within a day of arrival. Renaudot gradually added other functions to the Bureau: people could also inform the office if they had something to sell, and it acted as a pawn shop, with the aim of helping poor people raise money when they needed it. Renaudot and his medical friends also offered free medical consultations. In 1633 Renaudot began to publish an advertising newsletter called "Feuilles du

Bureau d'Adresse" which listed information under different headings: "Houses for sale in Paris," "Furniture for sale," "Offices requested for purchase," and so on. Each item had a number corresponding to its entry in the office's register, so that interested parties could quickly retrieve the details of the advertiser.

In addition, the Bureau became the hub of a conference series: the Conférences du Bureau d'Adresse, which ran from 1633 to 1642. The conferences were weekly educational events, open to all comers, at which participants discussed a vast range of topics, chiefly scientific and philosophical in nature: alchemy, the spread of disease, earthquakes, magnetism, talismans, witchcraft. They were so popular that attendance eventually had to be limited to one hundred people. Renaudot was a controversial figure in Paris because, like many other scientifically minded people of his day, he thought science ought to proceed on the basis of observation and experiment, rather than slavish adherence to the knowledge of the supposedly infallible ancients. He advocated a more scientific approach to medicine and, as a result, made enemies among the French medical establishment, which was still wedded to the traditional techniques of bloodletting. Similarly, at his weekly conferences, Renaudot insisted that participants should not give their names, but that their contributions should be considered on merit, rather than the reputation or social standing of the speaker. He wanted the conferences to be open to anyone, regardless of background. He thought science should be a collaborative process that proceeded through discussion and debate. Transcripts from his conferences were eventually published in five books, each nearly one thousand pages long, presented in a question-and-answer format, in which several speakers would address a particular question, such as "Were there more great men in any century before this one?"

None of Renaudot's pioneering attempts to encourage the circulation of information among the citizens of Paris lasted very long, however, with one exception. When his patron, Cardinal Richelieu, died in 1642, Renaudot had to give up his royal patent for the Bureau d'Adresse, and his enemies in the medical establishment managed to have him banned from practicing medicine. Renaudot died in 1653 in abject poverty ("as poor as a painter," one of his enemies sneered). He was survived by only one of his projects: the *Gazette de France,* a newspaper he had founded in 1631. Yet of all Renaudot's various projects, it was by far the least innovative and most conformist. Initially it consisted of translated stories taken from foreign gazettes and corantos. But to prevent other printers from issuing pirated copies of it, the *Gazette* was granted a monopoly giving it exclusive rights to cover foreign news and politics. This suited the French government, because it could control coverage of such matters more easily if there was only one place to find it. It also provided a convenient outlet for official announcements of the goings-on at court. The *Gazette* became the state newspaper, and its monopoly ensured that it was still going strong a century later. Unlike in Britain and America, where controls on the press had been loosened and the principle of freedom of expression had taken root in the political culture, the media system in eighteenth-century France was still subject to close scrutiny and tight control. In response to these restrictions, the media had by this time split into three distinct parts: the official, state-controlled publications such as the *Gazette de France*, the extraterritorial press, and the underground media.

Officially sanctioned publications, such as the *Gazette*, had an exclusive license, called a *privilège*, to cover a particular subject: while the *Gazette de France* covered foreign and political news, for example, the *Journal des Sçavans* covered science and academia. Dozens of other titles were produced for women, army officers, poets, theatergoers, and so on. The privilège system protected French publishers from competition and ensured that they only printed information that was favorable to the government, because if they did not, they might lose their privilège. As well as being licensed by the king, official periodicals were also subject to preapproval by censors, without which they could not be delivered using the royal mail system. The government employed nearly two hundred censors, and a special branch of the police existed to enforce their rulings and protect owners of privilèges from unauthorized printing on their turf. A wide range of official publications was available, in other words, but all of them studiously avoided discussion of domestic politics and religion.

The tedious columns of the *Gazette de France* were filled with reports of proclamations and ceremonies at court. One French writer joked that the *Gazette* only ever printed stories about the lottery, births in the royal family, and services in the Versailles chapel. This was not a million miles from the truth. And even when it did cover important political stories, the *Gazette* did so in a dry and uninformative manner. Consider, for example, how it reported the court drama of 1749, when Jean-Frédéric Phélypeaux, who was the comte de

Maurepas and one of Louis XV's most senior ministers, was suddenly dismissed by *lettre de cachet*—a royal letter, carrying the king's seal, that issued a ruling against which there was no appeal. The report appeared nearly two weeks after the events it described, and gave no indication of the earthquake in court politics caused by Maurepas's downfall:

> Paris, April 24. Toward 9 in the morning, the comte d'Argenson, Minister and Secretary of State in the War Department, went, at the King's order, to the residence of the comte de Maurepas and handed him a lettre de cachet by which His Majesty ordered him to retire to Bourges.

Despite its official monopoly, however, the *Gazette de France* was not the only game in town when it came to political news. Readers could also turn to the extraterritorial press, the French-language newspapers compiled and printed abroad and then imported to France. Being produced abroad meant they were, oddly enough, in a better position to write about domestic French politics, because they were unlicensed and did not have to worry about losing a privilège. The most prominent example was the *Gazette de Leyde*, published in the Dutch town of Leiden. Other extraterritorial French papers were produced in Cologne, London, and Avignon (which was not, at the time, part of France). These papers provided more detail than the *Gazette de France* about politics, for example by naming the ministers who were actually calling the shots, rather than pretending (as the *Gazette de France* did) that the king, who was the embodiment of the state, was personally responsible for all policy making.

These foreign papers were officially banned but were, in practice, tolerated by the authorities. Although they appeared to provide uncensored news, they were subject to control by the French state in various ways. They relied on postal distribution, for one thing, so if they went too far they could be excluded from the mail, as the *Gazette de Leyde* found when it tried to report on the attempted abolition of the *parlements* in the early 1770s. (The parlements were regional courts that were trying to maintain their influence as the monarchy became increasingly absolutist.) Correspondents and editors of extraterritorial papers could also be harassed or bribed, both at home and abroad, and papers often printed retractions in order to maintain official favor. As John Adams noted while acting as the American envoy in France, "all these papers . . . discover a perpetual complaisance for the French ministry, because . . . if an offensive paragraph appears, the entrance and distribution of the gazette may be stopped by an order from court." They were not as independent as they looked, in short, but the slightly freer extraterritorial papers had their uses, even to the government. In particular, they provided summaries in French of the most noteworthy stories in the foreign papers, which was quite helpful (Louis XV himself was said to read the *Gazette de Leyde*). The *Courrier de l'Europe*, produced in London in the 1770s, listed the news under the heading "Paragraphs extracted from English newspapers" and boasted that it offered "faithful extracts from the 53 gazettes that appear each week in London." It was briefly banned from the French mail in 1776 for running a letter from Paris attacking the government, but it was reinstated because it was such a valued source of news from Britain and America. The foreign papers could also be used by the French government to release information unofficially, for example to gauge the reaction to a proposed change of policy.

Even so, this was all a far cry from the vituperative, varied, and often vicious media environments in Britain, America, the Netherlands, or Germany. Amid such tight control, the real action took place in the underground social-media environment. Anyone who wanted to know what was really going on would turn to this third part of the French media system: the overlapping informal networks of gossip, songs, poems written on scraps of paper, materials printed with hidden presses, and handwritten news sheets called *nouvelles à la main* (which literally means "news by hand"). As Pierre Manuel, a political writer, put it in 1791:

> A people that wants to be informed cannot be satisfied with the *Gazette de France*. Why should it care if the king has performed the ritual of foot-washing for some poor folk whose feet weren't even dirty? Or if the queen celebrated Easter in company with the comte d'Artois? Or if Monsieur deigned to accept the dedication of a book that he may never read? Or if the Parlement, dressed in ceremonial attire, harangued the baby dauphin, who was dressed in swaddling clothes? The people want to know everything that is actually done and said in the court—why and for whom the cardinal de Rohan should have taken it into his head to play games with a pearl necklace; if it is

true that the comtesse Diane appoints the generals of the army and the comtesse Jule the bishops; how many Saint Louis medals the minister of war allotted to his mistress for distribution as New Year's presents. It was the sharp-witted authors of clandestine gazettes who spread the word about this kind of scandal.

## THE UNDERGROUND MEDIA

The demand for other sources of information increased during the 1740s as the French monarchy lurched from one crisis to another and became increasingly unpopular. Behind-the-scenes gossip could connect the dots of official information and provide a fuller and more credible picture. As one chief of police put it at the time, "The Parisians had more of a propensity to believe the malicious rumors and *libelles* that circulated clandestinely than the facts printed and published by order or with the permission of the government." Such gossip, passed from one person to another, circulated from the top of society to the bottom in a mixture of spoken, written, and printed forms.

The fall of the comte de Maurepas, which was so sparingly reported in the *Gazette de France*, was a direct consequence of his attempt to manipulate this underground media system. Maurepas was known for collecting satirical poems and songs, and he occasionally wrote them himself. There was no clear line between poems and songs; poems were simply easier to memorize when put to a familiar tune, and could then be spoken or sung. As they passed from person to person, whether orally or written on small scraps of paper, these ditties would be modified and reworked, with new verses added or names changed. "Qu'une bâtarde de catin," for example, was a popular song that functioned as a sort of "sung newspaper." Each of its verses satirized a different public figure, with the refrain blaming King Louis XV for letting things get into such a sorry state.

*Ah, there he is. Ah, here he is.*
*The one who doesn't have a care.*

Such poems could easily be updated in response to the news, a process of collective authorship that assimilated and encapsulated public opinion. Accessible even to the illiterate, they allowed information to travel freely between oral and written form, and also crossed class boundaries, traveling from the court to the street and back again. Maurepas and other courtiers would write or rewrite witty rhymes containing morsels of gossip, and these would circulate in court and then filter out via salons and coffeehouses to society at large. At the same time, rhymes being sung on street corners could reach the highest levels of the court. Maurepas, who among other things was in charge of the Paris police, collected songs and poems, many of them provided by informers, and wrote them in a *chansonnier*, or songbook, that eventually ran to thirty-five volumes. The king would sometimes ask him to recount the latest examples, both for entertainment and as a means of keeping track of public opinion. Which courtiers were being satirized, and what were the latest rumors about the royal family? The answers were to be found in the most popular circulating rhymes. Looking back on this period from the 1780s, one French writer recalled "the eagerness of the public to seek out these pieces, to learn them by heart, to communicate them to one another."

Maurepas's fall from grace in 1749 led to a series of arrests that illuminate the social-media system through which poems and anecdotes were propagated. His downfall was precipitated by an apparently innocent ditty about flowers.

*By your noble and free manner,*
*Iris, you enchant our hearts.*
*On our path you strew flowers,*
*But they are white flowers.*

Set to a popular tune, this rhyme began to do the rounds in the court at Versailles the day after Maurepas had attended a private dinner with the king, the king's mistress Madame de Pompadour, and her cousin, Madame d'Estrades. At the dinner, Pompadour had presented each of her three dining companions with white hyacinths from a bouquet. This provided an opportunity for a lurid pun: "white flowers" was also slang for the symptoms of venereal disease. The rhyme, in short, was a grave insult to Pompadour. But who had written it?

Pompadour was already the subject of dozens of insulting rhymes known as *poissonnades*, a reference to her maiden name, Poisson. She had repeatedly asked Maurepas, who oversaw the Paris police, to do more to suppress their circulation. But Pompadour was widely resented for her influence over the king, both within the court and beyond, and Maurepas preferred instead to permit and even encourage the flow of poissionnades in an effort to get Louis to renounce her. Pompadour, meanwhile, distrusted Maurepas and wanted Louis to replace him with one of her allies, the comte d'Argenson. When the "white flowers" song began to circulate it was immediately apparent that Maurepas, who had been one of only four people present at the dinner and was known for his love of satirical songs, must have written it. Louis sided with his mistress rather than his minister, and Maurepas was banished. The lettre de cachet dismissing him was delivered by Pompadour's friend, the comte d'Argenson, who then took Maurepas's place.

In the weeks after Maurepas's downfall a surge of new poissonnades began to circulate, including a rhyme entitled "The Exile of Monsieur de Maurepas" that called Louis a "monster" with a "black fury." At the king's behest, d'Argenson set out to find and punish its author. He told the police, who duly told their informants, that a reward would be paid to anyone who could provide information that might lead to the identification of the author of the offending rhyme. At the end of June the police received a note from an informant. In return for twelve gold coins (a year's wages for an unskilled laborer) he produced a copy of the rhyme and explained that it had been provided by François Bonis, a medical student. D'Argenson ordered his immediate arrest, and Bonis was tricked into boarding a carriage, which took him straight to the notorious Bastille prison. The account of his interrogation survives, recorded in the form of a dialogue.

> Asked if it isn't true that he composed some poetry against the king and that he read it to various persons.
>
> Replied that he is not at all a poet and has never composed any poems against anyone, but that about three weeks ago when he was in the hospital visiting abbé Grisson, the hospital director, at about four o'clock in the afternoon, a priest arrived . . . and that this priest, saying someone had had the malignity to compose some satirical verse against the king, pulled out a poem against His Majesty from which the respondent made a copy.

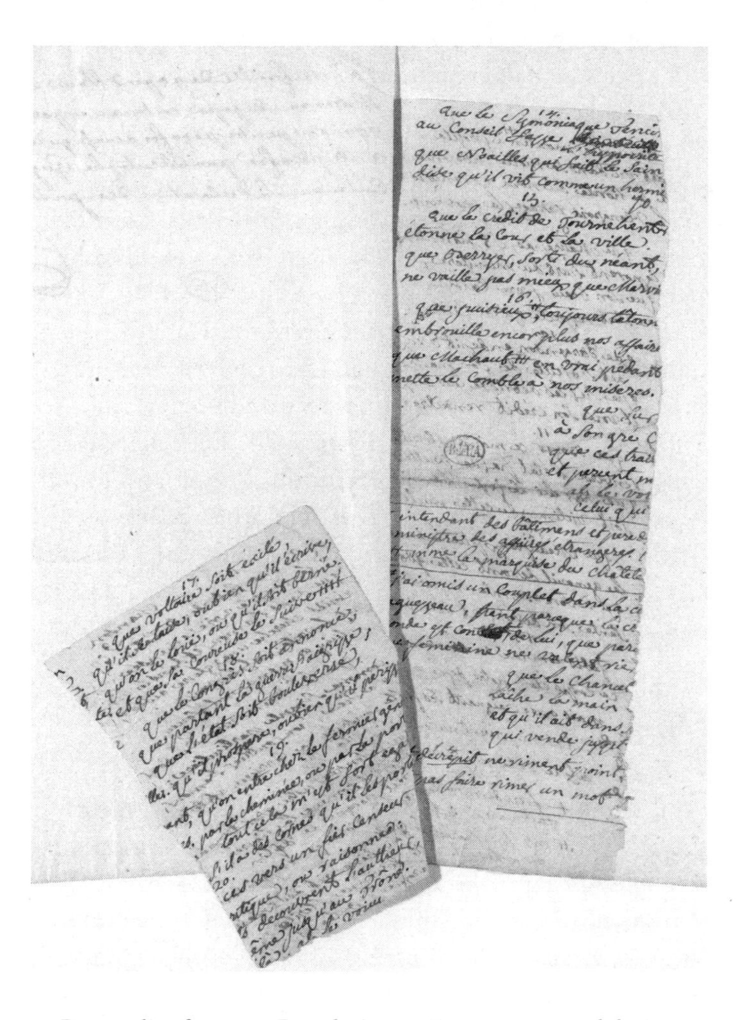

Poetry slips from pre-Revolutionary France, recovered during an interrogation at the Bastille in 1749. *Bibliothèque Nationale de France*

This was how rhymes circulated: in the form of small slips of paper that could be easily slipped into a sleeve or pocket. Many of them survive in the archives of the Paris police, who would search anyone they arrested for incriminating evidence. In cafés or private groups, people would trade rhymes and gossipy anecdotes, reading out the latest examples they had collected and, if particularly struck by a new rhyme they had not heard before, scribbling it down in order to be able to pass it on to others. Bonis did not know the name of the priest from whom he had copied the rhyme. So the police made him write a letter to his friend Grisson, to trick him into revealing the priest's name. The priest was duly arrested, as was a second priest from whom he, in turn, had copied the poem.

The police followed the trail, arresting one person after another, hoping to find the original author. The second priest had heard the poem from a third priest, who had heard it from a law student, who had heard it from a clerk, who had heard it from a philosophy student, who had heard it from a classmate, who had heard it from another student who could not be found. Jacques Marie Hallaire, the philosophy student in this lengthy chain, had also received three other poems from another priest, who had in turn received them from three other people. By the time the trail had run cold the police had arrested fourteen people for sharing the offending rhyme and found themselves trying to track the progress of five other poems as well. The investigation came to be known as "The Affair of the Fourteen" as a result.

It is not as surprising as it might seem that each person who was arrested was prepared to say who had supplied him with the poem. The police made it clear that anyone who could not explain where they had got it from would be assumed to be the author, and merely passing on a poem was a lesser crime than actually having written it. The search for a single author may have been misguided, however, because the rhymes and songs that circulated were commonly modified by the people through whose hands they passed—either deliberately, to improve them, or because of the imperfections of memory. Records of the interrogations of the fourteen suspects reveal the variety of settings in which poems and rhymes were shared and exchanged. Hallaire, the philosophy student, had copied down the offending poem at a dinner with some friends at his father's house. One of the priests involved had written the poem down after hearing it read aloud at a college dinner. Poems were changing hands almost everywhere.

For some people, known as *nouvellistes* or *bulletinistes*, trading rhymes and anecdotes written on small slips of paper was a full-time occupation. They compiled these circulating snippets into manuscript newsletters that they sent to subscribers who wanted to keep up with the latest gossip. They insisted this was harmless—merely passing on *bruits publics* ("public noises," or rumors) that anyone could hear just by walking into a coffeehouse—but they were considered dangerous by the state because their manuscript letters, unlike printed newspapers, could not be easily monitored or suppressed. These letters were then copied and passed along correspondence networks linking city, court, and provinces, even reaching foreign subscribers. One government minister complained to the chief of police that "experience has convinced us that of all classes of writers, the paid nouvelliste is the most difficult to restrain. What prudent man would put his faith in the conduct of a newsletter writer who calculates his profits according to the number of secret anecdotes he can gather?"

The pockets of Mathieu-François Pidansat de Mairobert, a nouvelliste who was arrested in 1749 for collecting and distributing scandalous poems about Louis XV and his mistress in coffeehouses, were full of slips of paper covered with poems, anecdotes, and morsels of gossip—a day's worth of gathering raw material. Mairobert was not connected directly to any of the fourteen suspects, though many of the same poems passed through his hands. He enjoyed spreading gossip so much that he would sometimes put slips of paper into his friends' pockets or leave them lying around in parks or cafés. A search of his apartment revealed another sixty-eight poems, reports, and anecdotes. But that was nothing compared with the apartment of Barthélemy-François Moufle d'Angerville, another nouvelliste, which produced seven boxes full of paper slips. Yet another manuscript newsletter was that compiled in the salon of Madame Doublet, who gathered her well-connected friends once a week to collect information. One of her servants would draw up two lists in large registers: one of credible news stories and one of gossipy anecdotes. The members of the salon would pass the registers around and read each story in turn, adding any further information they could provide. The final reports were copied and sent as a newsletter to Madame Doublet's friends, who shared it, in turn, with their friends. One recipient of her newsletter began paying scribes to make large numbers of copies, for sale to paying subscribers in the provinces. Others followed suit, ensuring that gossip from Paris spread far and wide.

Suppressing the circulation of manuscript newsletters was difficult, since they resembled ordinary letters. So the police instead tried to identify and imprison the most critical nouvellistes or, where possible, buy their favor. One nouvelliste, Charles de Fieux, also acted as a spy for the police, though eventually he fell foul of them. The police themselves also produced manuscript gazettes based on written reports from their spies, who differed rather little, if at all, from nouvellistes. It must have seemed as though the whole of Paris was passing around little slips of paper in a series of overlapping information networks, and compiling them into newsletters. An anecdote or rhyme could quickly travel from the top of society to the bottom and back again. As one observer complained at the time,

> A vile courtier puts these infamies into rhyming couplets and, through the intermediary of flunkies, distributes them all the way to the marketplace. From the markets they reach artisans, who in turn transmit them back to the noblemen who first wrought them and who, without wasting a minute, go to the royal chambers in Versailles and whisper from ear to ear in a tone of consummate hypocrisy, "Have you read them? Here they are. This is what is circulating among the common people in Paris."

Perhaps the most elaborate means by which gossip circulated was the roman à clef (literally, a "novel with a key"), a book that recounted the goings on in the French court in the form of a fictional account of a far-off land. "The Amours of Zeokinizul, King of the Kofirans," published secretly in 1747, is outwardly the tale of King Zeokinizul and his three mistresses, Liamil, Leutinemil, and Lenertoula, two of whom were pawns of the wicked courtier Kam de Kelirieu. Such was the king's love for Lenertoula that he took her with him when he went to war. But he fell ill and only recovered after confessing his sins and renouncing his mistress. To readers in the 1740s, this was obviously the story of their own king (Louis XV, pronounced "Louis Quinze") and his mistresses Madame de Mailly, Madame de Vintimille, and Madame de La Tournelle, two of whom were under the influence of the Duc de Richelieu. Other books dressed up the same story in slightly different disguises. "Tanastès," written by Marie Madeleine Joseph Bonafon, a twenty-eight-year-old chambermaid at Versailles, set the tale in a magical fairyland. "Secret Memoirs for a History of Persia," probably written by Madame de Vieuxmaison, moved the story to the Middle East. A police report described the author as "very clever and . . . very wicked, she writes poems and couplets against everyone . . . Her circle . . . is the most dangerous in Paris and is strongly suspected of having produced the ['Secret Memoirs']." Reading these books involved teasing out the many hidden meanings and references, and surviving copies often contain printed or written lists, or keys, that explain which real-life figure corresponds to each of the characters. Printed and distributed in secret, and passed from one friend to another, such books were difficult to suppress because they did not rely on postal delivery. And controls on printing weakened in the 1770s and 1780s in any case, as members of the ruling class found it useful to be able to issue defamatory pamphlets, or libelles, against each other.

For Maurepas, tolerating or even encouraging the circulation of gossip made sense, because it provided a means of monitoring and (via bribery and planting stories) influencing public opinion. This is reminiscent of the Chinese government's current approach to social-media censorship, which Min Jiang, a researcher at the University of North Carolina at Charlotte, calls "authoritarian deliberation." The authorities tolerate discussion within certain limits, providing a safety valve for discontent and advance warning of potential unrest, and only step in to censor sensitive topics or direct criticism of prominent members of the regime (as d'Argenson did in the case of the "Affair of the Fourteen"). The government also pays thousands of people, known as the Fifty Cent Party, to make posts or comments in support of government policies or actions, or to steer the conversation away from particular topics, in an effort to shape public opinion. China has constructed the most sophisticated scheme for monitoring and censoring social media in history, but it is unclear whether this situation is sustainable. In eighteenth-century France, relentless criticism steadily corroded respect for the monarchy and undermined the king's authority, though the process took decades. Whether in the form of poems, rhymes, songs, anecdotes, manuscript newsletters, or elaborate fantasy stories, much of the circulating gossip had a common theme: the corruption and immorality of the royal family and the malign influence of mistresses and senior courtiers. The French monarchy was unusually vulnerable to such personal attacks, built as it was on a cult of personality. But attacks on individual figures in the 1740s gradually gave way to attacks on the corruption of the system as a whole in the 1780s. In the words of Robert Darnton, a historian who has studied the media system of pre-Revolutionary France in detail, "the media knit themselves together in a communication system so powerful that it proved to be decisive in the collapse of the regime."

## "EVERY HOUR PRODUCES SOMETHING NEW"

By the late 1780s France was in crisis, having amassed an enormous public debt, around one third of it arising from its support of the American Revolution. Unable to raise taxes any further on commoners and unwilling to cut its own extravagant spending, the government decided instead to increase taxes on the clergy and nobility, who paid very little. But the nobles objected and refused to implement the new laws in the local provincial courts, or parlements, which they controlled. They insisted that a meeting of the three estates—the clergy, the nobility, and the commoners, collectively known as the Estates-General—was necessary to approve the new tax laws. Facing bankruptcy, the government had no choice but to agree, and in the summer of 1788 it lifted censorship restrictions to facilitate political discussion in the run-up to the meeting the following year.

The easing of censorship resulted in an immediate flood of pamphlets. More than fifteen hundred individual titles had appeared by the end of 1788, and another twenty-six hundred in the first four months of 1789, as the representatives, or deputies, for the three estates were elected. The total number of pamphlets printed in the year through May 1789 probably exceeded ten million copies. The result was an unprecedented national debate. Pamphlets made proposals for systems of finance and government, and praised or denounced political opponents. Both educated people and ordinary folk offered their reflections on how the government should be overhauled. Among the torrent of political pamphlets were several on press freedom, including a translation of Milton's *Areopagitica* by Honoré, Comte de Mirabeau, a politician and pamphleteer who was friends with Benjamin Franklin and Thomas Jefferson. He added a long introduction in which he attacked supporters of censorship, insisted that press freedom was a right, and pointed to its beneficial effects in Britain and America. Other writers advanced similar arguments, insisting that press freedom was necessary if deputies at the Estates-General were to be fully aware of their constituents' demands. A royal censor, meanwhile, issued a pamphlet insisting on the state's right to use censorship to maintain public order. But the official in charge of the book trade, Poitevin de Maissemy, admitted in May 1789 that "things have reached such a pitch that it is now almost impossible to prevent the circulation of these pamphlets." The English writer Arthur Young, who was traveling in France at the time, remarked after a visit to the "pamphlet shops" of Paris in June 1789 that "every hour produces something new. Thirteen came out today, sixteen yesterday and ninety-two last week."

Once the meeting of the Estates-General began, several papers sprang up to report on it, notably Jacques Pierre Brissot's *Patriote français* and Mirabeau's *Journal des Etats-Généraux*. Both Brissot and Mirabeau were familiar with the role the press had played in the American Revolution and were outspoken advocates of press freedom. Having learned about the effect of *Common Sense* in swaying public opinion, Brissot concluded that its success had depended on the way it had been reprinted and discussed in the newspapers, bringing it to a wide audience. He declared that "without newspapers the American Revolution would never have succeeded." But the new French papers were immediately banned by the government, which insisted that aside from its licensed journals, no publication could report on the Estates-General unless it had been directly authorized by the deputies. Brissot responded by publishing a new pamphlet calling for press freedom, and Mirabeau, who despite being a nobleman had been elected as a deputy to the third estate, hit upon the ruse of publishing his reports on the proceedings as letter-pamphlets entitled "The letters of the Comte de Mirabeau to his constituents." He was, he argued, just printing the letters he wrote to his constituents to keep them informed. Several other deputies followed suit. But they all offered dry summaries of debates, without explanation or commentary. For people who wanted to know what was really going on, the *Gazette de Leyde* still outdid them all. There was no need to bother with any of these new papers, it confidently told its readers, because they only "give false impressions of everything that has happened."

The remaining restrictions on the press did not last long, however. Rather than swiftly approving the new tax laws, as the government had hoped, the Estates-General immediately got bogged down in an argument about its own structure and powers, prompting the third estate, together with some sympathetic nobles and the majority of the clergy, to break away and rename itself the National Assembly. Concern that the king would use troops to shut down this assembly brought people out onto the streets of Paris in protest, and on July 14 crowds stormed the hated Bastille prison and a new municipal government was established in Paris. In the following weeks towns and cities across the country followed the lead of Paris, setting up municipal governments led by elected magistrates rather than officials appointed by the king. The authority of the old government collapsed, and with it the system of press regulation. Privileges and monopolies, along with any attempt to enforce censorship, vanished. The result was a free-for-all. Brissot relaunched his *Patriote français* on July 28, under the slogan "A free press is the eternally vigilant sentinel of the people." It was just one of dozens of new periodicals: more than one hundred and forty were launched in Paris in 1789, and another three hundred and thirty-five in 1790, though few of these survived more than one or two issues.

The media landscape, like the political scene, had been completely transformed in the space of just a few months. By the end of 1789 there was a huge choice of pamphlets and periodicals, offering a wide range of political viewpoints and reporting styles. Some offered sober commentary on the debates in the National Assembly: the weekly *Feuille villageoise*, for example, was aimed at rural readers, providing background information and explanations of the issues being discussed. As in America many periodicals, notably *L'Observateur*, gave abundant space to letters from readers, providing a platform from which they could respond to others' views and express their own. The distinction between a newspaper and a pamphlet was a fluid one, however. Many publishers of periodicals issued pamphlets as supplements, and if a particular issue of a newspaper did well it might be reissued as a pamphlet. And the more opinionated periodicals contained little by way of news and more closely resembled a series of polemical pamphlets. Camille Desmoulins wrote a series of polemics calling for a republic, rather than the British-style constitutional monarchy favored by Mirabeau and other moderates. The *Ami du peuple*, written by Jean-Paul Marat, offered a series of rabble-rousing speeches, rather than reports of recent events. He denounced particular deputies he disagreed with and accused them of covert support for the aristocracy or of having failed to represent their constituents properly. The assembly was mercilessly criticized in the press for being self-serving, hopelessly partisan, and incapable of getting anything done. Marat hated the assembly because it was not radical enough, while royalists hated it for failing to show sufficient deference to the king. Within the assembly itself, deputies complained that the unrestrained press had become too powerful and too critical, and should be reined in. But the few attempts to impose new controls on the press were ineffective. A free press was, after all, one of the most tangible and visible accomplishments of the revolution.

Fierce competition, together with a constantly shifting political environment, meant that few periodicals lasted very long. At any one time during the revolution there were typically twenty-five to thirty healthy titles in France, with many more close on their heels. To complicate matters further for publishers, paper and ink were hard to come by and skilled printers moved constantly from one printing shop to another in search of better pay. Printers were required to work through the night to prepare the next morning's papers. So many copies were needed—some papers were selling several thousand copies a day—that multiple presses had to be dedicated to printing the same publication in parallel, because each press could only turn out around a thousand copies a day. The *Journal du Soir* needed five presses, the *Chronique de Paris* needed seven, and the *Gazette universelle* needed as many as ten. Two rival publications called the *Ami du roi* fought for royalist readers after the editor and publisher fell out. There were half a dozen rival publications called *Père Duchesne*, written in the voice of a folk character who commented on the news in plain and vulgar language, each claiming to be the original and denouncing the others as impostors. Journalists savaged each other in print: Desmoulins' antiroyalist paper, the *Révolutions de France et de Brabant*, which he referred to as a "hypercritical journal," waged a constant war of words with its royalist rival, the *Actes des Apôtres*. Publishers who found themselves on the wrong side of the constant power struggles between revolutionary factions might suddenly have their presses confiscated or destroyed. As the struggle between the various revolutionary factions became increasingly violent, many writers and publishers were guillotined. Being a publisher during the French Revolution was often exhilarating, always precarious, and sometimes fatal.

## DOES A FREE PRESS LEAD TO FREEDOM?

Even in the midst of this chaotic environment, there was enormous optimism about the potential of a free press to enable a national conversation and promote democracy and freedom. Earlier in the eighteenth century French political theorists such as the Baron de Montesquieu and Jean-Jacques Rousseau had argued that true democracy was possible only in small cities, as in Ancient Greece, where the all citizens could gather in one place for discussion and collective decision making. But the more open media environment in Britain, and the effective use of the press during the American Revolution, suggested that printing offered a way to make democracy work at a national level, by allowing opinions to be shared and discussion to proceed over a wider area. As the French philosopher Nicolas de Condorcet put it, "the knowledge of printing makes it possible for modern constitutions to reach a perfection that they could not otherwise achieve. In this way a sparsely populated people in a large territory can now be as free as the residents of a small city . . . It is through the printing press alone that discussion among a great people can truly be one." Printing, in other words, led to freedom.

This view was widely shared among publishers and politicians of the time. Only with a free press, wrote Brissot, "could a large and populous modern country recreate the public forums of the classic city states; only through the press could one teach the same truth at the same moment to millions of men and only through the press could men discuss it without tumult, decide calmly and give their opinion." Brissot regarded the press as "the great tribune of humanity," a great instrument for distilling and giving voice to public opinion and thereby ensuring that politicians in the assembly could remain in tune with the people who had elected them. Similarly, Mirabeau wrote that "newspapers establish communications that cannot fail to produce a harmony of sentiments of opinion, of plans and of action that constitutes the real public force, the true safeguard of the nation." Newspapers and pamphlets would allow the twenty-eight million people of France, contended one writer at the National Assembly in 1790, to "be virtually present at the sessions of this august Senate, as if they were attending in person." These were the words of Étienne Lehodey de Saultchevreuil, founder of the *Journal logographique*, the aim of which was to report absolutely everything said in the assembly. The press, he believed, could relay the doings of government back to the people, and the will of the people back to their elected government.

Some even claimed that the outburst of pamphlets in 1788–89 meant printing had been responsible for the revolution itself and would lead to the inevitable spread of its ideals of freedom and equality, which were encapsulated in the Declaration of the Rights of Man issued by the National Assembly in 1789. "Blessed be the inventor of the printing press," declared the political writer Louis-Charles de Lavicomterie. "It is to him that we owe this wondrous revolution." Brissot wrote that if only everyone in the world spoke French, "the press would soon spread the French Revolution everywhere."

Brissot, Condorcet, and Lavicomterie were members of a group of intellectuals who set out to promote these ideals by demonstrating just how effectively the press could promote discussion and synthesize opinion by enabling information sharing within a large community. In 1790 they were among the founders of an organization called the Cercle Social ("social circle"). By holding public meetings and publishing newsletters and pamphlets, the Cercle Social aimed to provide a broad and inclusive platform for political conversation. On the door of its headquarters in Paris was a postbox, the "bouche de fer" ("iron mouth") into which letters, memos, and speeches could be posted by anyone who had an opinion they wished to express. It took its name from the metallic sculpture of a lion on its front, through whose mouth documents were posted. These would then be printed in the Cercle Social's newsletter, the *Bulletin de la Bouche de fer*, along with letters from its readers all over France and summaries of speeches made at the organization's public meetings. These meetings were held on Friday evenings in Paris and were attended by hundreds and in some cases thousands of people.

"The principle aim of the Cercle Social," one of its founders declared, was "to give the voice of the people all of its force." The Cercle Social formed links with other political clubs throughout France and tried to set up affiliates in several cities beyond Paris. It set out to provide a forum for political discussion accessible to anyone in the country, and for a while it succeeded: the *Bulletin de la Bouche de fer* became one of France's leading national newspapers for a few months in early 1791. There were even attempts to set up branches of the Cercle Social in other countries, to create a worldwide network. "The Cercle Social is not a club . . . it is an association of citizens spread all over the globe," declared its newsletter in February 1791. The Cercle Social also printed pamphlets written by some of its leading members, along with French translations of the works of Thomas Paine, who was by this time living in Paris—in Condorcet's house, in fact. But despite the club's inclusive rhetoric, most of its participants were well-to-do intellectuals, and the Cercle Social's grandiose experiment came to an end in July 1791 amid the political crisis triggered by Louis XVI's failed attempt to flee the country. Martial law was imposed and the democratic and republican ideals espoused by the Cercle Social came to be regarded as dangerously radical by the conservative authorities.

The most optimistic statement of the view that the sharing of information, facilitated by printing, would ultimately allow truth and freedom to prevail was written by Condorcet, in his *Outlines of an Historical View of the Progress of the Human Mind*. This epic work depicted human history as a narrative of gradual and triumphant progress toward economic and political freedom. Its optimism, which brings to mind the utopian claims made about "Internet freedom" today, is particularly striking because Condorcet wrote it while in hiding. By 1793 the democratic republicanism espoused by the members of the Cercle Social's leadership, which closely overlapped with the liberal Girondin faction, was regarded as insufficiently radical by the newly dominant Montagnard faction, led by Maximilien de Robespierre. Although he was a republican, Condorcet had been branded a traitor for arguing, unsuccessfully, that the king should merely be sentenced to forced labor rather than being executed.

His personal circumstances suggested otherwise, but Condorcet insisted that a free press was the guarantor of personal freedom. In *Outlines*, he remarked how fortunate it was that the liberating power of the press had not been recognized when it was first invented, "since priests and kings would infallibly have united to stifle, from its birth, the enemy that was to unmask their hypocrisy, and hurl them from their thrones." Once the genie was out of the bottle, however,

> the press multiplies indefinitely, and at a small expense, copies of any work . . . facts and discoveries not only acquire a more extensive publicity, but acquire it also in a shorter space of time . . . What was read before by a few individuals only, might now be perused by a whole people . . . A public opinion is formed, powerful by the number of those who share in it, energetic, because the motives that determine it act upon all minds at once, though at considerable distances from each other. A tribunal is erected in favor of reason and justice, independent of all human power, from the penetration of which it is

> difficult to conceal any thing, from whose verdict there is no escape . . . Every new error is resisted from its birth; frequently attacked before it has disseminated itself, it has not time to take root in the mind.

As for long-held misconceptions and errors, Condorcet wrote, "it is now impossible to prevent their discussion, impossible to conceal that they are capable of being examined and rejected, impossible they should withstand the progress of truths which, daily acquiring new light, must conclude at last with displaying all the absurdity of such errors. . . . In short, is it not the press that has freed the instruction of the people from every political and religious chain?" And trying to stamp out an inconvenient truth was rendered almost impossible, because only one printed copy, in a part of the world equipped with printing presses capable of reproducing it, was needed to ensure its survival.

> A single corner of the earth free to commit their leaves to the press, would be a sufficient security. How admidst that variety of productions, amidst that multitude of existing copies of the same book, amidst impressions continually renewed, will it be possible to shut so closely all the doors of truth, as to leave no opening, no crack or crevice by which it may enter? If it was difficult even when the business was to destroy a few copies only of a manuscript, to prevent for ever its revival, when it was sufficient to proscribe a truth, or opinion, for a certain number of years to devote it to eternal oblivion, is not this difficulty now rendered impossible, when it would require a vigilance incessantly occupied, and an activity that should never slumber?

As he wrote these words, however, Condorcet was surrounded by evidence of the drawbacks of an unfettered press in a chaotic, lawless environment. By 1793 the writers of pamphlets and newspapers were no longer engaged in political debate, but had descended to calling outright for the execution of their political opponents. Marat was put on trial for inciting violence against the Girondins, but he was acquitted of charges, leading to his assassination by a Girondin sympathizer. Meanwhile increasingly lurid and hateful "biographies" of Louis XVI, his queen Marie Antoinette, and various noblemen were doing the rounds, charging them with hampering the revolution. The duc d'Orléans was accused of trying to exploit the turmoil to install himself as regent in place of the king; like Louis, he was eventually guillotined. Other nobles were said to be plotting with foreign powers to invade France, or planning coups at home. The queen was accused of taking various revolutionary leaders as lovers in order to manipulate them and was said to be planning to blow up the National Assembly with a mine, set fire to Paris, and take the throne for herself. Some of these accounts provided "evidence" in the form of letters supposedly written by the people in question, in which they conveniently incriminated themselves.

The subtext was the same in each case: counterrevolutionary conspiracies orchestrated by people in high places threatened to rob the people of their hard-won liberties. Marie Antoinette in particular was depicted as being behind all the setbacks that had befallen the revolution: "All our past, present and future calamities have always been and will always be uniquely her work," declared one of the many accounts of her life. Amid the twists and turns of the revolution, this was a comforting explanation, intended to pave the way for her eventual execution. Similar character assassinations were produced to justify the executions of various revolutionary figures, in some cases after their deaths. Robespierre and Brissot were honored in this way, as was Mirabeau, even though he died of illness in 1791. All were depicted as evil super-villains manipulating events to their own advantage before eventually being undone by honest revolutionaries. An interesting shift occurs in these libelles, compared with those produced in the prerevolutionary period: instead of attacking their subjects using ridicule, the tone is one of simple denunciation; and rather than using examples of sexual escapades to emphasize their subjects' decadence and immorality the emphasis is on financial improprieties: theft, fraud, refusal to pay debts, and schemes for self-enrichment.

Condorcet himself escaped the guillotine, but only because he died in prison under mysterious circumstances in 1794, possibly after taking poison. His fate embodies the gap between the French revolutionaries' optimism about the power of the press to promote freedom, at least in theory, and the actual use of the press to condemn and denounce in practice. John Adams pointed this out in a series of notes in the margins of his copy of Condorcet's *Outline*. When Condorcet asserted that the press established "a new kind of authority . . . from which is exercised a less tyrannical empire over the passions but a more firm and lasting power over reason," Adams noted that "the empire of the press, over the passions, in the hands of Marat and others was more tyrannical than the government of Cesar Borgia." Condorcet suggested that "every new error is resisted from its birth," but Adams responded that "there has been more error propagated from the press in the last ten years than in a hundred years before." In the case of the French Revolution, at least, the press had promoted tyranny and falsehood just as effectively as freedom and truth.

In retrospect, the problem was that revolutionary leaders such as Mirabeau, Brissot, and Condorcet assumed that the removal of controls on printing would allow the inherent but unspoken unity of public opinion to express itself. But instead the press revealed and amplified the differences between the various revolutionary factions. In America the free press had helped make apparent an underlying broad consensus in favor of independence, but in France it highlighted the lack of agreement about what should replace the monarchy. The French revolutionaries assumed a stable country was a united one. They did not imagine a future of competing political parties, as emerged in America in the 1790s, accompanied by a fiercely partisan press. Instead they saw their opponents as enemies who had to be destroyed, and the press as a powerful tool in achieving that aim. The press was ultimately tamed and institutionalized with the rise to power of Napoleon, who is said to have remarked that he had more to fear from four hostile newspapers than from a thousand bayonets. A free society requires a free media system, but the lesson of the French Revolution is that a free media system on its own will not necessarily result in a free society.

# CHAPTER 9

# THE RISE OF MASS MEDIA: THE CENTRALIZATION BEGINS

*The world fashioned by the mass media is a public sphere in appearance only.*
—Jürgen Habermas

## MASS-PRODUCED NEWS

IN THE EARLY hours of November 29, 1814, the printing staff at the *Times*, London's leading newspaper, waited anxiously by their idle presses. The paper's proprietor, John Walter, had told them that he was expecting important news from Europe, where the major powers were in the process of redrawing the map after the defeat of Napoleon Bonaparte in May that year. The *Times* had established a reputation for being first with the news from war-ravaged Europe, often learning of developments before the British government itself. So it was not unusual for Walter to delay the printing of the paper to allow the latest reports from Europe to be included at the last minute.

Four centuries had passed since the invention of the printing press, but the *Times*, like other newspapers, was printed using hand presses that had changed very little since Gutenberg's time. The frame that secured the press and the screw mechanism that pressed the type onto the paper were, by the early nineteenth century, both made of metal rather than wood, but otherwise the design would have looked familiar to a printer from the fifteenth century. A team of skilled workers could operate such a press at a rate of two hundred and fifty to three hundred impressions per hour, and the *Times* had several presses that could be operated in parallel. Each night, working flat out, its printers could produce five thousand copies of the newspaper in two or three hours, to ensure they were ready for distribution at dawn. But as the minutes ticked by, the news from Europe failed to arrive, and no order was issued to start printing the paper without it.

Then, at six o'clock in the morning, Walter strode into the print room holding a copy of that day's paper. It had, he explained to his astonished workers, already been printed using steam-powered machinery that he had secretly had installed in the basement of another building. This subterfuge had been necessary because Walter's first steam press had been demolished by printers who were concerned, like other manual workers at the time, that they would be replaced by machines. Walter told the printers that he would not be swayed from his decision to adopt steam printing and assured them that he would ensure that they did not suffer as a result. The vastly more efficient steam press, devised by a German printer named Frederick Koenig, could produce around fifteen hundred impressions per hour. As the *Times* declared in that morning's edition, "Our journal of this day presents to the public the practical result of the greatest improvement connected with printing, since the discovery of the art itself. The reader of this paragraph now holds in his hand one of the many thousand impressions of the *Times* newspaper which were taken off last night by a mechanical apparatus." With steam printing, information had become an industrial product.

In theory, this meant newspapers could now be produced in large quantities at a lower cost, allowing the *Times* to drop its cover price and increase its circulation. In practice, however, the price of the *Times* and other British newspapers was kept artificially high by the stamp duty, which had taken over from censorship as the British government's preferred means of limiting the flow of information and trying to ensure that news was available only to the wealthy elite. The duty was introduced in 1712 at a rate of a halfpenny a sheet, was gradually increased during the eighteenth century, and reached four pence in 1815. This raised the cost of the new, steam-printed *Times* from three pence to seven pence. Given that a laborer at the time earned an average of thirty-four pence a day, a daily newspaper was beyond the means of the most people, even though a growing proportion of them could read. The stamp duty was condemned by publishers as a "tax on knowledge," and some radical newspapers were produced illegally, without paying it. Publishers in Britain were to spend the next four decades waging a campaign against the stamp duty.

The situation was very different in the United States, where a free press, making information as widely available as possible, was regarded as an important pillar of the new democracy—so much so that it was enshrined in the Bill of Rights. As Thomas Jefferson famously put it in 1787 in a letter to a friend, "Were it left to me to decide whether we should have a government without newspapers, or newspapers without government, I should not hesitate a moment to prefer the latter." An attempt to introduce a stamp tax on newspapers in Massachusetts in 1785 caused outrage, reviving memories of the hated British stamp tax, and it was quickly repealed. The idealism of the 1780s evaporated during the 1790s, however, as dozens of new papers sprang up, espousing fiercely politically partisan views.

John Adams, who had defeated Thomas Jefferson to become president in 1796, became so exasperated at the attacks on him in Republican newspapers that he passed the Sedition Act, making it a crime to defame or criticize the president or his government. Adams's attempted clampdown on press freedom was also prompted, in part, by concerns that too much press freedom, as in France, might contribute to political instability. But it backfired. Imprisoned journalists became heroes—one was even elected to Congress while in jail—and in 1800 Adams was defeated by Jefferson. But Jefferson's staunch advocacy of press freedom wavered when he found himself, in turn, on the receiving end of partisan attacks. "Nothing can now be believed which is seen in a newspaper," he lamented to a friend in 1805. Jefferson recognized the difficulty of regulating the press without undermining its freedom, however, and ultimately maintained his position that a free press, for all its faults, did more good than harm. "Where the press is free, and every man able to read, all is safe," he wrote to a friend in 1816. The attempts to curtail press freedom in the 1790s and 1800s actually strengthened it, establishing the right to criticize the administration and supporting vigorous and lively political debate.

Moreover, rather than imposing taxes on newspapers, as the British did, the American authorities actively subsidized them by giving them discounted postage rates and special privileges. The Post Office Act of

1792 set low rates for postal delivery of newspapers and enshrined the right of newspaper editors to exchange papers with each other for free. Postmasters also helped out by collecting newspaper-subscription fees from readers. And America's government-run postal network was far more extensive than the delivery networks in other countries: by 1828 there were seventy-four post offices per hundred thousand people in America, compared with seventeen in Britain and four in France. By 1832 newspapers accounted for 95 percent of all mail by weight, but only 15 percent of revenue, and per-capita newspaper circulation in America was more than double that in Britain. If anything, government support for newspapers was too generous. By 1843 America's 1,634 newspapers were exchanging seven million free copies a year—which meant that, on average, each newspaper editor received twelve free newspapers a day. Postmasters-general repeatedly complained that this was a waste of effort, because many of these newspapers were simply discarded.

Even though the *Times* of London was the first newspaper to adopt a steam press, it was in America's far more open media environment that a new model first emerged, based on reducing cover prices and increasing circulation. It was pioneered by the New York *Sun*, a paper launched by Benjamin Day, a twenty-two-year-old printer, on September 3, 1833. At the time daily newspapers in America typically sold for six cents and were mostly sold by subscription, paid as an annual fee. This put daily papers beyond the means of most people and kept their circulations small: American papers sold an average of twelve hundred copies a day in 1830.

The *Sun*, with its motto "It shines for all," was different. It sold at the knockdown price of one cent. But selling a paper at such a low price was only financially sustainable if it could attract a large readership and then bring in lots of money from advertisers. "The object of this paper," Day announced, "is to lay before the public, at a price within the means of every one, all the news of the day, and at the same time offer an advantageous medium for advertisements." The two objectives were mutually reinforcing and interdependent. Getting the ball rolling was difficult, because advertisers would not sign up if there were no readers, and a low price and large circulation depended on having plenty of advertising revenue. During 1833 there were several false starts as different publishers tried to launch penny newspapers. Day jump-started his paper by reprinting the advertisements from the established six-penny newspapers in the *Sun*, and then signing up advertisers for real.

He also paid hawkers to sell the paper on the streets, borrowing a technique already used in London. Newsboys had to pay sixty-seven cents for one hundred copies of his paper, which they could then sell for a penny each. They were not allowed to return unsold copies, so they had a strong incentive to sell the paper aggressively. Day, meanwhile, did not have to worry about making a loss on unsold copies, and readers did not have to sign up for a whole year's subscription but could buy individual copies when they wanted to. A further innovation was that advertisers were compelled to pay in advance, which also insulated the publisher from risk. The *Sun* was an instant success and by January 1834 it had a readership of five thousand, making it the biggest paper in America. Previously the best-selling paper had been the *Courier and Enquirer*, a New York business paper with a circulation of twenty-five hundred.

The *Sun*'s broad appeal did not rest simply on its low price and easy availability. It was also the result of its down-to-earth coverage, with an emphasis on anecdotes, morality tales, crime reports, quirky news items, and human-interest stories designed to appeal to ordinary people. This was a marked change from the usual newspaper fare of political stories and speeches, business items, and shipping news. Accounts of tiger hunting in India jostled in the *Sun*'s columns with reports of local fires and riots and a roundup of theories about the origin of the world. In 1835 the paper ran a series of sensational reports on the discovery of winged creatures on the moon, which had supposedly been spotted by John Herschel, a famous astronomer who had built a large telescope in Cape Town, South Africa. Herschel and his telescope were real, but the inhabitants of the moon were entirely fictional. The hoax was eventually exposed, but not before it had helped increase the *Sun*'s readership.

As the *Sun*'s circulation grew Day raised his advertising rates and in 1835 he bought a steam press, which allowed him to expand the paper's circulation to fifteen thousand. By 1838 circulation had reached thirty-eight thousand, making the *Sun* the most popular newspaper in the world. Day's success prompted others to emulate his approach. Some thirty-five new penny papers were launched in New York alone between 1834 and 1839, only a handful of which survived very long. Penny papers also sprang up in Boston, Philadelphia, and Baltimore.

In each case their circulation soon dwarfed that of the existing papers in each city. The *New York Herald*, launched by James Gordon Bennett in 1835, was selling twenty thousand copies a day within two years of its launch, despite an increase in its cover price to two cents. The *New York Times* launched in 1851 as a penny paper.

Here, for the first time, was a truly mass medium: mass-produced news for a mass audience. Bennett declared that his paper was "equally intended for the great masses of the community—the merchant, mechanic, working people—the private family as well as the public hotel—the journeyman and his employer—the clerk and his principal." But it depended on more than just the steam press and the new advertising-supported business model. Improvements in papermaking technology also helped make newspapers more affordable. Wider adoption of a process called stereotyping meant there was no need to set identical type to print multiple copies of a paper in parallel; instead type could be set once and then used to make lead molds from which multiple identical metal plates could be cast. These technological factors increased supply, while social and political changes increased demand. In America, growing literacy rates and the expansion of political rights to all adult white men in the 1820s meant that newspapers were no longer the sole preserve of the political and business elite. Wider literacy created the market for the penny papers, but they in turn accelerated the advance of literacy by providing ordinary people with something they wanted to read.

The established papers grumbled that the new penny papers carried vulgar stories and immoral advertising. Papers had previously been picky about their advertisers, refusing on moral grounds to carry advertisements for lotteries, theaters, or businesses that opened on Sundays, for example. The penny papers, by contrast, would print advertisements from anyone who paid: their advertising columns were dominated in particular by patent medicines of dubious efficacy. Some papers were criticized for running advertisements for abortionists. Proprietors of penny-press papers insisted that running any advertisement, provided it was not offensive or misleading, was part of their mission to make information more widely available than it had been in the past. As the *Boston Daily Times* put it,

> Some of our readers complain of the great number of patent medicines advertised in this paper. To this complaint we can only reply

> that it is for our interest to insert such advertisements as are not indecent or improper in their language . . . It is sufficient for our purpose that the advertisements are paid for, and that, while we reserve the right of excluding such as are improper to be read, to the advertising public we are impartial, and show no respect to persons, or to the various kinds of business that fill up this little world of ours. One man has as good a right as another to have his wares, his goods, his panaceas, his profession, published to the world in a newspaper, provided he pays for it.

## THE RISE OF THE REPORTERS

One consequence of this new mass-media model was that there was more money available to pay for newsgathering. The idea that newspapers ought to employ people to seek out stories for publication seems obvious now, but it was still a novelty in the early nineteenth century. In Britain, newspapers began to carry detailed reports of parliamentary debates in the 1770s, and the men who wrote them would appear to have been the first dedicated reporters. Because the taking of notes was not permitted, the main qualification for the job was to have a prodigious memory. William "Memory" Woodfall, the printer and editor of the *Morning Chronicle*, earned his nickname from his astonishing ability to recall and summarize hours of debate into a detailed and accurate report. Eventually, in 1783, the rules were changed to permit note-taking, which made life much easier for parliamentary reporters. By the beginning of the nineteenth century London newspapers were also starting to send reporters to noteworthy trials and even, in a couple of cases, to battlefields in Europe.

Such zeal for newsgathering appeared slightly later in America, where early nineteenth-century newspapers still relied on letters, speeches, excerpts from pamphlets, poetry, and stories taken from other papers to fill their columns. "No mail yesterday," grumbled the *New Orleans Gazette* in 1805. "We hardly know what we shall fill our paper with that will have the appearance of news." A writer in the *Weekly Herald*, recalling the 1820s, noted that "the newspapers of that day relied altogether upon their exchanges for news, and, of course, the intelligence which they gave the readers was meagre, stale and unsatisfactory." One editor filled his empty columns by reprinting the obituary of

George Washington from a few years earlier, or chunks of the Bible. "The editor of this paper," declared the *Democratic Press* of Philadelphia in 1814, "will sincerely thank any of his friends, who will favor him with loan of foreign papers, particularly French or English." In 1816 New York's two most popular daily newspapers contained, on average, fewer than two columns of actual news.

Things began to change in the late 1820s as some papers began to employ political correspondents in Washington, D.C., and others made more of an effort when it came to business news. In particular, two rival New York papers, the *Journal of Commerce* and the *Courier and Enquirer*, began to use pony expresses to deliver news from other cities, and fast boats to meet incoming vessels and get foreign news a few hours early. With the advent of penny papers, which had to be sufficiently compelling to get people to buy them every day, competition for timely or original news stories intensified. Bennett, founder of the *New York Herald*, agreed to pay one of his sources five hundred dollars for every hour by which he beat other papers in getting news from Europe.

From the very first issue of the New York *Sun* in 1833, Benjamin Day realized that an endless stream of human-interest stories could be found in the courts. He paid a British writer, George Wisner, to attend the police court in lower Manhattan and write darkly humorous accounts of the proceedings, such as this report from the *Sun*'s inaugural issue: "John McMan, brought up for whipping Juda McMan, his darling wife—excuse was that his head was rather thick, in consequence of taking a wee drap of whiskey. Not being able to find bail, he was accommodated with a room in Bridewell [the city jail]." Wisner, who also helped with the *Sun*'s typesetting, was paid four dollars a week, plus a share of the paper's profits. He had learned his trade from John Wight, a writer for the London *Morning Herald* who is considered the world's first dedicated crime reporter, and who published a book of his crime reports in 1824. But rather than mocking the criminality of the working classes for the entertainment of wealthier readers, as Wight did, the New York penny papers delighted in providing the lurid details of thefts and murders, a more sensationalist approach with mass appeal. Bennett justified his paper's devotion to crime reporting on the basis that it was educational. "A good police reporter in such a city as New York is a useful person to society," he wrote. "An exact and correct record of crime—ingenious crime, not

# THE SUN.

NUMBER 16.] NEW YORK, FRIDAY, SEPTEMBER 20, 1833. [PRICE ONE PENNY.

**PUBLISHED DAILY,**

AT 222 WILLIAM ST.............BENJ. H. DAY, PRINTER.

The object of this paper is to lay before the public, at a price within the means of every one, ALL THE NEWS OF THE DAY, and at the same time afford an advantageous medium for advertising. The sheet will be enlarged, as soon as the increase of advertisements require it—the price remaining the same.

Yearly advertisers, (without the paper,) Thirty Dollars per annum—Casual advertising, at the usual prices charged by the city papers.

☞ Subscriptions will be received, if paid in advance, at the rate of Three Dollars per annum.

PASSAGE FROM ENGLAND AND IRELAND.—Persons wishing to engage passages from the ports of Liverpool, Belfast, Londonderry, Dublin, Sligo, and Cork, can at all times have them secured in first class packet ships, by applying to
A. THOMPSON & CO. or to
s2 AB'M BELL & CO. 33 Pine st.

FOR CHARLESTON.—To sail 20th instant. The very superior new copper fastened and coppered ship Harriet & Jessie, S. W. McKown, commander, having the greater part of her cargo engaged, will sail as above. For freight which will be taken low, or passage, apply on board at Rector st. wharf, or to
s19 TINKHAM & HART, 97 Pine st.

WANTED—A good schooner of about 130 tons, for the southern trade. Apply to
Fowle, Dunham & Day, 59 Pearl street.

TO LET.—A new two story brick House, in Waverly Place, late 6th street, next to the corner of Grove street, opposite the Greenwich Dispensary, just finished in handsome style, and every way suitable for a genteel family. Rent low to a good tenant, and will be let till 1st MAY, or for a longer period. Immediate possession can be had. Apply on the premises, or at 241 Water st. s12

TO LET—A new two story House, No. 105 Sullivan st., finished in the best manner, and well calculated for the accommodation of a respectable family. Apply to Mrs. Glover, 109 Laurens st. or Mr. Adams, Fulton Bank. s12

CAPS AND STOCKS—At Harding's Cap and Stock Factory, No 130 Chatham street, opposite Rosevelt street—Stocks made on superior bristle frames, of the following materials, viz:
Satin, Trimmed, Pleated, and Plain.
Bombazine, do do do
Silk Velvet do do do
White and colored Hair Cloth, do
Also—A large assortment of Broadcloth, Haircloth, Fur, Silk, Merino, and Circassian Caps, of the latest fashions, and of superior make, wholesale and retail.
s14 R. L. HARDING.

PROPOSALS FOR COAL.—Sealed proposals will be received by the Commissioners of the Alms House, until the 5th September next ensuing, at their office in the Park, for one hundred tons best Schuylkill Coal, to be delivered on or before the last day of October next. s2

H. ROWELL & CO, 40 Canal street, near Broadway, New York, Manufacturer of Paste and Liquid Blacking, of a superior quality, warranted not to injure the Leather, which they offer upon the most liberal terms.

☞ H. R. & Co. pledge themselves to give the utmost satisfaction, and should the article not prove as good as is warranted, the purchaser is at liberty to return and receive the cash for the same.

Also,—Black and Red Ink, of the first quality, and furnished by the wholesale. s17

181 CHATHAM-ST. OPPOSITE THE CHAPEL.

DELISSA & CO. begs to inform the ladies and gentlemen of New York and its environs, that they have just arrived from Europe with a most splendid assortment of French, British, and German Fancy Articles; Jewelry, Mosaic and Composition Gold of all descriptions; Musical Instruments in all varieties; Ladies Work Boxes with and without music, Gentlemen's Dressing Cases, &c. which D. & CO. are enabled to dispose of at the most reasonable terms, at fixed prices only—occular demonstrations will convince the public, that their way of dealing will give satisfaction to all who feel inclined to patronize them. Great allowance to retail dealers. s14 tf

THE CHEAPEST Ready Made LINEN STORE, DENAUX'S, No. 181 Hudson street. s21

WANTED.—3 or 4 girls to learn the Tayloring business in its most Fashionable Style, by a first rate Tayloress. References as regards respectability &c., will be required and given. Address W. at this office.

J. VAN DUSER, 212 Division street, Cordial Distiller, dealer in Liquor, Lemmon Syrup, &c. s14

JOB PRINTING neatly executed at this office s14

SPECTRE OF THE BROCKEN.

We copy the following interesting account of this extraordinary aerial phenomenon from "Brewster's Letters on Natural Magic."

The Brocken is the name of the loftiest of the Hartz Mountains, a picturesque range which lies in the kingdom of Hanover. It is elevated 3300 feet above the sea, and commands the view of a plain seventy leagues in extent, occupying nearly the 200th part of the whole of Europe, and animated with a population of above five millions of inhabitants. From the earliest periods of authentic history, the Brocken has been the seat of the marvellous. On its summits are still seen huge blocks of granite, called Socerer's Chair and the Altar. A spring of pure water is known by the name of the Magic Fountain, and the anemone of the Brocken is distinguished by the title of the Socerer's Flower. These names are supposed to have originated in the rites of the great idol Cortho, whom the Saxons worshipped in secret on the summit of the Brocken, when Christianity was extending her benignant sway over the subjacent plains.

As the locality of these idolatrous rites, the Brocken must have been much frequented, and we can scarcely doubt that the spectre which now so often haunts it at sunrise must have been observed from the earliest times; but it is nowhere mentioned that this phenomenon was in any way associated with the objects of their idolatrous worship. One of the best accounts of the spectre of the Brocken is that which is given by M. Haue, who saw it on the 23d of May, 1797. After having been on the summit of the mountain no less than thirty times, he had at last the good fortune of witnessing the object of his curiosity. The sun rose about four o'clock in the morning through a serene atmosphere. In the south-west, towards Achtermanshohe, a brisk west wind carried before it the transparent vapors, which had not yet been condensed into thick heavy clouds. About a quarter past four he went towards the inn, and looked round to see whether the atmosphere would afford him a free prospect towards the southwest, when he observed at a very great distance, towards Achtermannshohe, a human figure of a monstrous size. His hat, having been almost carried away by a violent gust of wind, he suddenly raised his hand to his forehead to protect his hat, and the colossal figure did the same. He immediately made another movement by bending his body—an action which was repeated by the spectral figure. M. Haue was desirous of making further experiments, but the figure disappeared. He remained, however, in the same position, expecting its return; and in a few minutes it again made its appearance on the Achtermannshohe, when it mimicked his gestures as before. He then called the landlord of the inn, and having both taken the same position which he had before, they looked towards the Achtermannshohe, but saw nothing. In a very short space of time, however, two colossal figures were formed over the above eminence; and after bending their bodies and imitating the gestures of the two spectators, they disappeared. Retaining their position, and keeping their eyes still fixed upon the same spot, the two gigantic spectres again stood before them, and were joined by a third. Every movement that they made was imitated by the three figures, but the effect varied in its intensity, being sometimes weak and faint, and at other times strong and well defined.

In the year 1798 M. Jordan saw the same phenomenon at sunrise, and under similar circumstances, but with less distinctness, and without any duplication of the figures.

*Decision of Character.*—You may recollect the mention, in one of our conversations, of a young man who wasted in two or three years a large patrimony, in profligate revels with a number of worthless associates calling themselves his friends, till his means were exhausted, when they of course treated him with neglect or contempt. Reduced to absolute want, he one day went out of the house with an intention to put an end to his life; but wandering awhile almost unconsciously, he came to the brow of an eminence which overlooked what were lately his estates. Here he sat down, and remained fixed in thought a number of hours, at the end of which he sprang from the ground with a vehement exulting emotion. He had formed his resolution, which was, that all these estates should be his again; he had formed his plan too, which he instantly began to execute. He walked hastily forward, determined to seize the very first opportunity, of however humble a kind, to gain any money, though it were so despicable a trifle, and resolved absolutely not to spend, if he could help it, a farthing of whatever he might obtain. The first thing that drew his attention was a heap of coals shot out of carts on the pavement before a house. He offered himself to shovel or wheel them into the place where they were to be laid, and was employed. He received a few pence for the labor; and then, in pursuance of the saving part of his plan, requested some small gratuity of meat and drink, which was given him. He then looked out for the next thing that might chance to offer; and went with indefatigable industry through a succession of servile employments, of longer or shorter duration, still scrupulously avoiding, as far as possible, the expense of a penny. He promptly seized every opportunity which could advance his design, without regarding the meanness of occupation or appearance. By this method he had gained, after a considerable time, money enough to purchase, in order to sell again, a few cattle, of which he had taken pains to understand the value. He speedily but cautiously turned his first gains into second advantages; retained without a single deviation his extreme parsimony; and thus advanced by degrees into larger transactions and incipient wealth. I did not hear, or have forgotten, the continued course of his life: but the final result was, that he more than recovered his lost possessions, and died an inveterate miser, worth £60,000. I have always recollected this as a signal instance, though in an unfortunate and ignoble direction, of decisive character, and of the extraordinary effect which, according to general laws, belongs to the strongest form of such a character. [Foster's Essays.

*The Morning Air.*—There is something in the morning air that, while it defies the penetration of our proud and shallow philosophy, adds brightness to the blood, freshness to life, and vigour to the whole frame—the freshness of the lip, by the way, is, according to Dr. Marshall Hall, one of the surest marks of health. If ye would be well, therefore—if ye would have your heart dancing gladly, like the April breeze, and your blood flowing like an April brook—up with the lark—'the merry lark,' as Shakspeare calls it, which is 'the ploughman's clock,' to warn him of the dawn;—up and breakfast on the morning air—fresh with the odour of budding flowers and all the fragrance of the maiden spring;—up from your nerve-destroying down bed, and from the foul air pent within your close drawn

The cover of the New York *Sun* from 1833. *Courtesy American Antiquarian Society* vulgar drunken brawls—is useful as a warning and a beacon for others to avoid."

Newspapers began to rely more heavily on personal observation and interviews by reporters, rather than on documents, as the basis of stories. Bennett broke new ground in 1836 when he went to a brothel where Helen Jewett, a twenty-three-year-old prostitute, had been murdered with an axe. He questioned the owner of the house and published a detailed description in the *New York Herald* of the crime scene, where the victim's body still lay:

> The countenance was calm and passionless. Not the slightest appearance of emotion was there. One arm lay over her bosom . . . For a few moments I was lost in admiration at this extraordinary sight—a beautiful female corpse—that surpassed the final statue of antiquity. I was recalled to her horrid destiny by seeing the dreadful bloody gashes on the right temple, which must have caused instantaneous dissolution.

It is unclear how truthful his account was, but Bennett's intrusion was remarkable even by the standards of later "doorstepping" journalists. He produced two more reports from the crime scene, describing what he found when he searched the victim's bed and writing desk in the first, and recounting a conversation with the matron of the brothel in the second. The veracity of these reports, like that of the crime-scene description, is uncertain; rival newspapers accused Bennett of fabricating them to sustain the surge in circulation his paper was enjoying as a result of its coverage of the case. Bennett mounted a long campaign to defend the chief suspect, Richard P. Robinson, perhaps because an open-and-shut case of a murder of a prostitute by one of her clients was less exciting, and would sell fewer newspapers, than the more elaborate theories advanced by the *Herald*. Robinson was eventually acquitted, despite strong evidence against him.

In 1834 the *New York Transcript*, a new penny paper launched soon after the *Sun*, observed that of the eleven traditional papers in the city, only one or two employed reporters to gather the news. But the *Transcript* and the *Sun*, it noted, each employed four reporters "exclusively to obtain the earliest, fullest, and most correct intelligence on every local incident; and two of these latter arise at 3 in the morning, at which hour they attend the police courts . . . while others are obtaining correct information about the city." Soon after its launch in 1835 the *New York Herald* had appointed full-time correspondents in Washington, Jamaica, and Key West, and in 1838 it hired six correspondents in Europe. The *New York Tribune*, founded by Horace Greeley in 1841, had a staff of fourteen local reporters by 1854, responsible for unearthing stories at police and fire stations, in the hospitals, lecture theaters, and law courts, and at the docks.

The paradoxical result was that although mass-circulation newspapers expanded the availability of information—the *Sun* boasted that "the penny press, by diffusing useful knowledge among the operative classes of society, is effecting the march of independence to a greater degree than any other mode of instruction"—the job of gathering material and setting the news agenda gradually became concentrated in the hands of a select group of editors, reporters, and proprietors. Previously, newspapers had consisted of musings from the proprietor publisher mixed with excerpts of notable speeches and pamphlets, letters from readers, and reports taken from other papers. This gave way, in the early nineteenth century, to a new model based around a permanent staff of writers. One survey of American newspapers found that the number of articles written by papers' own writers increased from 25 percent to 45 percent between the 1820s and 1850s.

Newspapers also began to distance themselves from their readers in another way. Under the new penny-press model, advertisers were a more important source of revenue than readers. From a business point of view, the goal of a newspaper was now to gather as large a readership as possible in order to provide an audience for advertisements. Readers were no longer seen as participants in a conversation taking place within the newspaper's pages; instead they had become purely consumers of information and, potentially, of the products and services offered by advertisers. Rather than providing a platform for the circulation and discussion of documents created by others, newspapers became vehicles for the delivery of their own reports.

## MORSE'S SOCIAL NETWORK

In March 1843 a painter-turned-inventor named Samuel Morse was awarded thirty thousand dollars by the United States government to demonstrate the viability of his new invention: the electric telegraph. Morse was by no means the first person to have built an electric telegraph, the basic principles of which had been understood by scientists for some decades. But he is remembered as its inventor today because he had two crucial advantages. The first was his refusal, from the day he first witnessed a demonstration of electromagnetism in 1832, to give up on his dream of building a long-distance messaging system. The second was the relatively simple design of his telegraph apparatus. His British rivals, William Cooke and Charles Wheatstone, had devised an electric telegraph that required five or six wires to be run between the two stations, which was expensive. Working with his assistant Alfred Vail, Morse had developed a far simpler system that ultimately required just a single wire.

Morse's telegraph shifted the complexity from the hardware into the software: characters were encoded used sequences of dots and dashes, now known as Morse code, though it was largely Vail's work. (It was Vail who had the idea of counting the letters in a print shop to determine their relative frequency, in order to make the assignment of combinations of dots and dashes to particular letters as efficient as possible. E, the most common letter in English, is a single dot in Morse code.) The use of this code made Morse's telegraph harder to operate, but much cheaper to build. Amid much skepticism from those who insisted that the whole enterprise was a fraud, Morse used the government funds to build an experimental line between Washington and Baltimore. On May 1, 1844, he demonstrated the power of his invention by telegraphing the result of the Whig National Convention, being held in Baltimore, to Washington. The news was verified when written confirmation arrived by steam train from Baltimore, sixty-four minutes later. Morse had shown that electricity could transmit information far more quickly than it could be physically carried by horse, ship, or train.

With backing from investors, Morse formed a company to build telegraph links between New York, Philadelphia, and Boston, the first of which (between New York and Philadelphia) opened for business in January 1846. The business community quickly recognized the value of the technology, prompting other firms to start building telegraph lines too. America's telegraph network expanded so quickly in the years that followed that it was almost impossible to keep track of its size. "No schedule of telegraphic lines can now be relied upon for a month in succession," complained one writer in 1848, "as hundreds of miles may be added in that space of time. It is anticipated that the whole of the populous parts of the United States will, within two or three years, be covered with net-work like a spider's web." By 1848 there were around two thousand miles of wire, and by 1850 there were more than twelve thousand miles operated by twenty different companies. "The telegraph system [in the United States] is carried to a greater extent than in any other part of the world," wrote the Superintendent of the Census in 1852, "and numerous lines are now in full operation for a net-work over the length and breadth of the land." By this stage there were more than twenty-three thousand miles of line in operation, with another ten thousand under construction. "No invention of modern times has extended its influence so rapidly as that of the electric telegraph," declared *Scientific American* in 1852. "The spread of the telegraph is about as wonderful a thing as the noble invention itself."

Morse's apparatus, it soon turned out, could be simplified even further. The initial design used long and short pulses of current, generated by pressing a telegraph key, to draw a pattern of dots and dashes on a paper tape at the other end of the line. These dots and dashes were then translated into a message. But skilled operators quickly learned to read incoming messages by ear, simply by listening to the rhythmic sound of the apparatus. Paper tapes were abandoned in favor of telegraph sounders, which made distinctive clicking noises as the current was switched on and off. The use of keys and sounders made possible direct, real-time person-to-person conversations over the wires. Telegraph operators could chat with each other by tapping on their keys and then listening to the answering clicking sounds. When a single wire connected several telegraph stations, all the operators along the line could hear everything that was transmitted and join in the unofficial banter, in effect occupying a single, shared chat room. Some could even recognize their friends merely from the style of their Morse code—something that was, to a really skilled telegrapher, as recognizable as an individual human voice.

Not every telegrapher was so skilled, however, so operators on a particular telegraph line would each choose a two-letter signature or "sig" with which to identify themselves. And rather than spelling out every word ("Philadelphia calling New York") letter by letter in laborious detail, conventions arose based on short abbreviations. There was no single standard: different dialects or customs arose on different telegraph lines. However, one listing of common abbreviations compiled in 1859 includes "I I" (dot dot, dot dot) for "I am ready," "G A" (dash dash dot, dot dash) for "go ahead," "S F D" for "stop for dinner," "G M" for "good morning," and so on. Numbers were also used as abbreviations: 1 meant "wait a moment," 2 meant "get answer immediately," and 33 meant "answer paid here," for example.

As well as chatting, exchanging gossip, and telling jokes, telegraph operators played chess and checkers over the wires during quiet periods when there were no messages to transmit. Telegraphers often made friends over the wires; romances between operators who met each other online were not unknown. Such was the sense of online camaraderie that some operators in remote places preferred to commune with their friends on the wires than with the local people. Thomas Stevens, a British telegraph operator stationed in Persia, shunned the local community in favor of telegraphic interaction with other Britons. "How companionable it was, that bit of civilization in a barbarous country," he wrote of his telegraphic friends, most of whom he had never met and were thousands of miles away.

On one occasion in the 1850s the employees of the American Telegraph Company lines between Boston and Calais, Maine, held a meeting by telegraph after hours to discuss the resignation of their superintendent. The meeting was attended by hundreds of operators in thirty-three offices along the seven-hundred-mile line. According to a contemporary account, "each speaker wrote with his key what he had to say, and all the offices upon the line received his remarks at the same moment, thus annihilating space and time, and bringing together the different parties, in effect, as near to each other as though they were in the same room, although actually separated by hundreds of miles." After passing various resolutions the meeting was adjourned "in great harmony and kindly feeling" after about an hour of discussion.

Telegraphers were members of the world's first online community, in instant contact with distant colleagues. For ordinary people who had to pay by the word to send messages, however, social banter via the telegraph was too expensive to be practical. Indeed, for society as a whole, rather than being a medium of direct personal interaction with others, the telegraph was quite the opposite, because it further centralized the impersonal flow of information.

American newspapers quickly embraced the new technology, with its ability to deliver timely news updates. But maintaining a network of distant correspondents who sent in reports via telegraph was expensive. Moreover, rival reporters sending dispatches from a particular telegraph office were all paying to send broadly similar reports, but each had to wait to use the line in turn. So in 1848 several New York newspapers teamed up to pool their resources, improve their bargaining position with the telegraph companies, and reduce their costs. The result was the Associated Press, a system that gradually expanded to include newspapers throughout America. By joining the AP, a newspaper gained access to a stream of news reports from its correspondents around the country and abroad. Florid, opinionated copy gave way to reports written in a dry, brief style (today we still call it "telegraphic") that presented facts plainly, in order of importance—the so-called "inverted pyramid" style. Such reports had the advantage that they could be printed in any newspaper, regardless of its political affiliation. This new, telegraphic writing style also influenced public speaking: short sound bites became popular because they were easier for stenographers to transcribe, and cheaper and quicker for reporters to transmit.

By the mid-1850s a typical American daily newspaper contained at least two columns of AP reports, and the AP came to wield enormous influence, particularly after free newspaper exchanges by post were abolished in 1873. Three other telegraphic news agencies agreed in 1870 to carve up the rest of the world: Reuters covered the British empire, Havas covered France and its colonial possessions in Africa and beyond, and a third agency, Wolff, covered Germany and its sphere of influence. The trio agreed not to compete but to share their reports, and they were later joined by the AP.

The telegraph increased the speed with which news could be delivered and made foreign news more widely available than ever before. Speed was everything; newspapers came to value timeliness at the expense of depth. Some people worried that the obsession with speed was not merely undesirable, but dangerous. In 1881 a New York doctor, George Beard, published a book called *American Nervousness*, which blamed the telegraph and the press for contributing to an epidemic of "nervousness" caused by the increase in the pace of business and social life. "The telegraph is a cause of nervousness the potency of which is little understood," he declared.

Writing in the *Atlantic Monthly* in 1891, W. J. Stillman, a journalist and critic, decried the way the telegraph had changed how information was distributed and consumed. "America has in fact transformed journalism from what it once was, the periodical expression of the thought of the time, the opportune record of the questions and answers of contemporary life, into an agency for collecting, condensing and assimilating the trivialities of the entire human existence," he moaned. All this was, he worried, changing the way people's brains worked. "The effect is disastrous, and affects the whole range of our mental activities. We develop hurry into a deliberate system, skimming of surfaces into a science, the pursuit of novelties and sensations into the normal business of our lives . . . The frantic haste with which we bolt everything we take, seconded by the eager wish of the journalist not to be a day behind his competitor, abolishes deliberation from judgment and sound digestion from our mental constitutions. We have no time to go below surfaces, and as a general thing no disposition."

In theory the telegraph had the potential to overcome the limits of geography and increase the direct flow of information between people in different places. But in practice it greatly concentrated and homogenized the flow of information—and left at least some people feeling dazed and overwhelmed. The possibility of using communications technology to create a shared social space was apparent only to the small community of telegraph operators who actually worked the wires. One telegrapher, looking back on his career in 1902, referred to the camaraderie among

> those who, in widely separated cities, are brought in daily touch by a wire used jointly by all. In idle intervals, on an Associated Press circuit, for example, a wire touching at a dozen or more cities distance is lost sight of, and all the features of personal intercourse are distinctly present. Stories are told, opinions exchanged, and laughs enjoyed, just as if the participants were sitting together at a club. They grow to know each other's habits, moods, and foibles, their likes and dislikes and when there is a break in the circle through the death of a member, his absence is felt just as in personal association.

## MARCH OF THE MOGULS

Successful penny papers were very profitable—Day made a profit of twenty thousand dollars in 1835—which gave their owners the cash to invest and expand. As the penny papers grew in circulation, the cost of launching a new paper steadily increased. Day started the *Sun* on a shoestring, and Bennett needed just five hundred dollars to found the *Herald* in 1835. Greeley launched the *Tribune* in 1841 with two thousand dollars in cash and one thousand dollars' worth of printing equipment. A decade later the cost of launching a new penny paper, the *New York Times*, had risen to seventy thousand dollars. In London the *Daily News*, edited by a fast-rising journalist and author named Charles Dickens, was launched in 1846 at a cost of one hundred thousand pounds. The launch of Britain's *Daily Mail* in 1896 cost five hundred thousand pounds. Newspaper publishing had turned into a centralized, capital-intensive industry.

Newspaper proprietors became powerful figures who wielded vast political influence over their readers and, in some cases, ran for office themselves. Horace Greeley, editor of the *New York Tribune*, helped found the short-lived Liberal Republican Party, which allied itself with the Democratic Party, and he was Democratic candidate for the presidency in 1872 (though he failed to win). The influence of proprietors increased as they began to assemble large publishing empires, linking up multiple newspapers by telegraph, allowing them to share reports with each other, and granting the proprietor national influence. This was the strategy followed by William Randolph Hearst and Joseph Pulitzer in America, Lord Northcliffe in Britain, and Leopold Ullstein and Rudolf Mosse in Germany.

Hearst was the epitome of this new breed: the media mogul. Having dropped out of Harvard, he learned about the newspaper business by taking a job at Pulitzer's New York paper, the *World*. Hearst then applied an exaggerated version of Pulitzer's approach—a combination of human-interest stories, scandal, crime reports, and tub-thumping campaigning on behalf of the working man—at the San Francisco *Examiner*, which he bought in 1887 using money from his father, who had made a fortune in mining. Having turned the *Examiner* around, Hearst bought the New York *Morning Journal* and initiated a vicious battle with Pulitzer that drove the circulation of individual newspapers above one million copies for the first time. The two tried to outdo each other in sensationalism, inventing stories and faking pictures in what came to be known as "yellow journalism."

Most famously, Hearst used his papers to stoke anti-Spanish sentiment in 1898, printing lurid accounts of Spanish persecution in Cuba and helping turn public opinion in favor of war with Spain. Hearst is supposed to have told one of his artists, who wished to return from Cuba because not much was happening, "You furnish the pictures, and I'll furnish the war." This tale is probably untrue, but there is no question that Hearst enjoyed wielding the power that his newspapers gave him. War finally broke out when an American ship, the USS *Maine,* exploded in Havana harbor. The explosion was probably the result of the accidental detonation of the ship's ammunition store, but the *Journal* declared it "the work of an enemy," America blamed the Spanish, and hostilities began. HOW DO YOU LIKE THE JOURNAL'S WAR? asked the headline in Hearst's paper.

Hearst also did his best to wield political power directly, serving in the U.S. House of Representatives, seeking the Democratic nomination for the presidency in 1904, and running for election as mayor, governor, and lieutenant-governor of New York, though without success. In part this was because his reputation was stained by the accusation that the strident journalism of his papers had played a role in the assassination of President William McKinley by an anarchist in September 1901. The *Journal* had attacked McKinley relentlessly, despite his successful prosecution of the war with Spain, and had run an editorial earlier that year stating that "if bad institutions and bad men can be got rid of only by killing, then the killing must be done." (Hearst insisted that the editorial had been published without his knowledge.) In 1906 Hearst was savaged in a speech made by the secretary of state, Elihu Root, who denounced him as "an insincere, self-seeking demagogue," effectively putting an end to Hearst's campaign for the governorship of New York. He ended up controlling thirty large American newspapers, amassed a valuable art collection, hobnobbed with celebrities and politicians, and was the model for Orson Welles's Citizen Kane.

The result was that newspapers at the end of the nineteenth century were very different beasts from those at the beginning. In 1800 most newspapers had been simple publications for which, in many cases, one person acted as editor, publisher, printer, and editorial writer, filling columns with petitions, speeches, letters from members of the community, and reports copied from other papers. Setting up a new press had required a modest investment by a merchant or entrepreneur. But by the second half of the nineteenth century newspapers were being written by dedicated reporters, supplied with foreign news by telegraph, funded primarily by advertisers, and printed on expensive steam presses. Readership was broader than ever, but the gulf between readers and writers, easily stepped over in the eighteenth century, grew ever wider as both ownership and authorship became steadily more concentrated.

This shift was lamented by the German philosopher Jürgen Habermas in his book *The Structural Transformation of the Public Sphere*, published in 1962. To Habermas, the advent of coffeehouses, salons, literary journals, and a free press in eighteenth-century Europe had constituted the emergence of a public sphere, providing spaces in which citizens could discuss matters freely in public, as social equals. These public spaces were, in theory, open to all (though participation was, in practice, skewed toward the intellectual elite, and women were almost entirely excluded) and were by and large politically independent. They could therefore offer a venue in which legal and political matters could be discussed and criticized, and they began to provide a check on the legitimacy of laws and policies. But as mass media emerged in the mid-nineteenth century, in the form of newspapers and magazines with large circulations, this fragile public sphere disintegrated, Habermas argued.

Over the course of the nineteenth century the newspaper evolved from being a handmade, local publication into the product of a vast, powerful, and lucrative industry. In the process its capacity to act as a platform for discussion within a community was greatly diminished, and it became almost exclusively a one-way medium. Large media organizations, controlled by and operated in the interests of a small group, came to dominate the media landscape, providing their owners with a powerful means to influence public attitudes. In the process, media became a product to be passively consumed, rather than an environment in which to actively participate. Yet despite the huge shift during the nineteenth century, an even bigger change was to come in the twentieth century—taking the centralization of the media environment still farther.

# CHAPTER 10

## THE OPPOSITE OF SOCIAL MEDIA: MEDIA IN THE BROADCAST ERA

*Wireless is a thrilling pastime. Fancy a boy sitting in his room at home with his fingers on a telegraph key and a telephone receiver to his ear . . . any boy can own a real wireless station, if he really wants to.*
—A. Frederick Collins, *The Book of Wireless*, 1915

### THE GOLDEN AGE OF WIRELESS

FOR AMERICAN BOYS growing up in the early twentieth century there was no better hobby, declared Hugo Gernsback in "A Sermon to Parents," than tinkering with the nascent technology of radio telegraphy. Admittedly, Gernsback, an entrepreneurial immigrant from Luxembourg, had a vested interest in promoting radio. A pioneer in amateur radio, he was the founder of the Electro Importing Company, also known as Telimco, which in 1905 began selling a "Complete Outfit" of equipment required for sending and receiving Morse code by radio for eight dollars and fifty cents. Gernsback's sermon appeared as an editorial in the Telimco catalogue and was intended to reassure parents who were worried about their child's interest in this strange new hobby. His opening argument was that experimenting with radio equipment would keep young boys at home, away from bad influences and out of trouble. "He doesn't want much, just something to dabble, to tinker, to experiment with and to keep his inborn insatiable curiosity satisfied," he wrote. "His workshop, his small electric laboratory or his Wireless Den are the most powerful home attractions for the 20th Century Boy." Better still, Gernsback argued, it was a hobby that would position a boy well for the future. "Electricity and Wireless are the coming, undreamed of, world-moving forces. Don't kill the electric spark in your boy. It costs little to keep it going, and some fine day it will pay you and your boy handsome dividends." Gernsback considered radio to be "the best possible foundation of the future self made man." No doubt he had in mind one radio enthusiast in particular, whose boyhood hobby had led to fame and fortune: Guglielmo Marconi.

As with the electric telegraph, the idea of radio telegraphy had been discussed in the scientific community for some years, but it took someone with a broader vision and entrepreneurial zeal to get things moving. Samuel Morse had played that catalytic role for the telegraph, and Marconi, a young Italian inventor, did the same for radio. Born in 1874 into a wealthy family, he was educated by private tutors and, in his teenage years, by Augusto Righi, a professor at the nearby University of Bologna. Righi taught Marconi about electromagnetic waves, which had been discovered by the German scientist Heinrich Hertz in 1886. Seized by the possibility of using these waves as a means of long-distance communication, Marconi embarked upon a series of experiments at his parents' house in the summer of 1894, in which he gradually improved Hertz's apparatus. One night in December that year he invited his mother into his attic workshop to show her the results: pressing a switch caused a bell to ring in an adjoining room, about thirty feet away, without the use of any connecting wires. Over the course of the following year he gradually extended the range of his equipment by refining its design and using larger antennae. He had soon moved the receiver outdoors into the garden, where it was manned by his older brother Alfonso, who would wave a flag to indicate that a signal had been received. By the end of the year Marconi could send radio signals more than a mile, and Alfonso was firing a gun to acknowledge successful reception.

Having failed to impress the Italian authorities with his invention, Marconi went to London, where he hoped it would arouse more interest. He won the backing of William Preece, the chief electrical engineer of the General Post Office, who arranged for him to demonstrate his invention to the British government. While in Britain

Marconi successfully sent Morse-code messages several miles over both land and water, and he gave two lectures in London outlining his achievements. He sent messages across the English Channel in 1899 and was invited to America that year by the *New York Herald* so that his equipment could be used to cover the America's Cup yacht race. In December 1901 he sent signals across the Atlantic for the first time, from Cornwall to Newfoundland. This was an impressive achievement, but the most immediate practical use for the technology was for ship-to-shore and ship-to-ship communication, rather than long-distance telegraphy. The Marconi Wireless Telegraph Company quickly established itself as the dominant provider of wireless equipment and services to both commercial and military shipping.

The new technology also attracted the interest of hobbyists and experimenters who wanted to try it for themselves, particularly in the United States, where there were no restrictions on the use of wireless equipment. At the dawn of the twentieth century, owning a wireless set opened a window on a new world. Amateurs could listen in on government transmissions and send messages to each other. Some of them had better equipment than the navy itself. Setting up a transmitter and receiver was not for the fainthearted, and communication required the use of Morse code. Even so, it became a popular hobby for boys, encouraged by Gernsback and other cheerleaders. In October 1908 *Electrician and Mechanic* magazine reported that "wireless telegraph mania" had broken out among the young men of Baltimore. "The ages of these amateurs vary greatly, the youngest being under fifteen years . . . at least thirty different wireless enthusiasts have excellent receiving and transmitting stations."

It was the same in other cities. "I pick up messages from all sorts of places," a young enthusiast in New York told *Modern Electrics* magazine in April 1909. "There must be hundreds of experimental stations in New York, and the messages are going through the air all night long. My apparatus has a radius of seventy-five miles." *Electrical World* estimated there to be "not less than 800 amateur wireless-telegraph stations in Chicago" in 1910. Its active radio community was organized around the Chicago Wireless Club, which sent out test messages every evening at eight o'clock to allow enthusiasts to test their equipment. Some members of the club had used kites to elevate their antenna wires and extend the range of their transmissions, *Electrical World* reported. *Modern Electrics* was one of two magazines established by Gernsback to cater to radio enthusiasts, the other being the *Electrical Experimenter.* In 1909 he also founded a national society for amateur radio fans, the Wireless Association of America, which quickly signed up thousands of members. In 1912 Gernsback put the number of "wireless experimenters and amateurs" in the United States at four hundred thousand. The majority of radio stations at the time were capable of both sending and receiving, though the sending range was usually much smaller than the receiving range. It was a vast conversation over the airwaves, open to anyone who wanted to join in.

All this was exciting for the participants, but the drawbacks of a complete free-for-all with thousands of radio stations soon became apparent. Early radio equipment broadcast on a wide range of frequencies simultaneously, so that transmissions would be picked up by all receivers within range. Despite the efforts of Marconi and others, the problem of tuning—limiting circuits to operating on particular radio frequencies, to eliminate interference with others—proved difficult to solve. As *Electrical World* observed in 1906, "The whole situation has become very complicated and unless something is done about it soon the effect will be like a telephone exchange with all the subscribers on the same line and all trying to talk at once . . . the time has now come when in wireless telegraphy it is either regulation or chaos, and of the two the former is certainly to be preferred." Concern grew as amateur stations began to interfere with commercial and naval transmitters, both accidentally and deliberately. *Electrical World* reported a case in 1907, involving an amateur operator who lived near the naval yard in Washington, D.C.: "A youth living near by, the son of a policeman, has set up a station of his own, and takes delight in interpolating messages during official exchanges. He has represented himself to be at distant naval stations or at sea on warships equipped with wireless apparatus. The local police authorities were appealed to, but said they had no power to interfere with the young man's experiments."

By 1908 the controversy over wireless interference, and the conflict between amateur enthusiasts and commercial and naval stations, had moved from specialist publications to the mainstream press. "STOP IT, KID! CRIES CONGRESS TO THE AMERICAN BOY," trumpeted an article in the *San Francisco Call* in March 1908. It depicted the situation as a fight between President Theodore Roosevelt and plucky youngsters "whose strenuosity, energy, curiosity, desire to know things, to do things and whose penchant for experimenting is so characteristic of the spirit of our best nation." The report noted the growing volume of complaints about interference in San Francisco, Washington, D.C., and other cities, which had led to calls for regulation. But most of the boys in and around San Francisco, the paper said, were behaving themselves.

> From sundown until midnight the ether is full of flitting aerograms . . . It is in the early hours of the evening, usually from 8 to 10 o'clock, that the boys around the bay are most busy in conversing with one another by means of their stations. And it is during that time that the "interferences" complained of usually occur. To the honor and credit of the boys it must be said I do not personally know of an instance where they made public any messages that they may have picked up, and some of them are quite competent operators. Most of them vigorously condemn the "breaking in" when the government or commercial stations are busy and refuse to answer with their spark at such times. One of them said to me, "Some of the boys are queering the game for us all, but the majority of them are sensible and stay out when the ships and stations around the bay are working."

But some charges against amateurs were more serious: they were accused of hindering marine rescue operations or broadcasting false distress signals. In January 1909 attempts to aid a crippled steamship were hampered by amateurs who broadcast several sets of false coordinates, sending rescuers on a wild goose chase. Captain K. W. Perry, who had led the rescue effort, told the *Christian Science Monitor* that the time had come for regulation. "We have long felt the necessity for some regulation in the use of wireless outfits, but the imperative need of such regulation has been demonstrated in the experiences of the past few days." Several more similar incidents followed, and when amateurs hindered the rescue of the *Terry*, a torpedo boat destroyer, in 1912, the government came under pressure to act. As a navy official told the *Electrician and Mechanic*,

> The incident of the *Terry* is argument enough for a federal license law . . . For more than an hour amateur operators interfered with the receipt of the message of distress. They were asked repeatedly to cease their activity in sending messages to each other. Instead of complying with the request, several of them retorted with impudent replies . . . Our country is the only one in the world in which all

> wireless operators are not required to have a license . . . We do not wish to be represented as discouraging young men who are ambitious in carrying on experiments in wireless operation. For the most part, they are young geniuses who have built their own stations. But when it is realized how serious their interference is at times, and what it might cost if some vessel was in distress, it can be appreciated that some action must be taken. The final solution lies in the federal license.

The sinking of the *Titanic* was the final straw. When the giant liner signaled that it had hit an iceberg, America's east coast lit up with speculation and rumor, flashed from one wireless station to another. Amateur wireless operators then became a convenient scapegoat for the *Titanic*'s owner, the White Star Line, as it struggled to explain what had happened. "From the offices of the White Star line came reports to the President that it was practically impossible to get any reliable information by wireless because of the great number of wireless concerns breaking into the field and because of the work of amateur operators," reported the *New York Herald*. "It appears that the disaster to the *Titanic* had no sooner been flashed over the seas than about every wireless instrument along the coast within range began operations, sending and receiving with no thought of others, so that the net result soon became a hopeless jumble, from which distorted and inaccurate messages were patched up in haphazard fashion and announced to the anxious world." President William Howard Taft called a special meeting at the White House to discuss the regulation of wireless. But, the *Herald*, reported, there was "little necessity for argument, for the staggering disaster of the *Titanic*, coupled with the wireless chaos concerning relief work and accurate reports, was in itself a demonstration which will serve for all time to show that regulation is imperative." The engineer in chief of the navy, Hutch Cone, told the *Herald* that "If there ever was a demonstration that regulation of wireless is necessary, this is it."

It subsequently transpired that the real problem was that the volume of outgoing telegrams being sent by the *Titanic*'s passengers was so great that the ship's wireless operator had failed to pass on a courtesy message from a nearby ship, the *Caledonian*, warning of ice in the area. Instead, he told the *Caledonian*'s operator to get off the air and stop blocking his transmissions. The *Caledonian* then switched off its radio and failed to hear the *Titanic*'s distress signal. The regulations passed in the wake of the *Titanic* disaster, the Radio Act of 1912, duly introduced new rules for the handling of distress signals by shipping, including the standardization of frequencies and the requirement that the radio be manned at all times. The regulations also introduced strict limits on amateur radio stations, requiring transmitters to be licensed, limiting them to wavelengths of two hundred meters and below, and imposing restrictions on the transmitting power, location, and operating hours of amateur stations. For the first time a clear line segregated amateurs from commercial and military users.

The act also gave the president the power to shut down radio stations "in time of war or public peril or disaster." When the First World War broke out in Europe, the navy was put in charge of ensuring that radio operators respected the neutrality of the United States, which it did in part by closing down amateur stations. When America entered the war in 1917, even using a receiver was banned, halting the use of radio equipment by amateurs altogether. When the ban was lifted in 1919, radio enthusiasts eagerly looked forward to picking up where they had left off. Technological improvements during the war—in particular, the advent of continuous-wave transmitters based on vacuum tubes—meant that wireless signals were no longer only dots and dashes of Morse code but could also carry sound, making possible radiotelephony. Hugo Gernsback, writing in the *Electrical Experimenter* under the headline AMATEUR RADIO RESTORED, declared:

> When we put our sets away two years ago we were accustomed to hear nothing but the crisp dots and dashes in flute-like, staccato sounds coming from the high power stations which we all had learned to love. But the war has changed everything—even radio, for now the radio telephone has come into its own. Where formerly there was nothing but the tah-de-dah in our phones, the ether is now filled with the human voice flung far and broad over the land—nay, over the oceans—and as the months roll by the dots and dashes will grow less and less, and the human voice will come in over our aërials more and more, which is as it should be . . . While no doubt many of us will still cling to the dots and dashes, the radio telephone will probably soon be used in overwhelming numbers.

Under the headline NOW THE BOY WIRELESS CAN GET BUSY AGAIN, the *Seattle Daily Times* looked forward to the rebirth of amateur radio. "No other country in the world may boast so large an army of amateur wireless men as America," it declared. "It was estimated before the war that there were at least 175,000 amateur stations scattered far and wide throughout the United States. Now that peace is assured the number will probably be even greater . . . With the increased facilities for long-distance work the American boy becomes in a sense a citizen of the world. From his home station, probably constructed in his study room, he enjoys a power which a generation ago would have seemed magical." Yet the switch from Morse code to audio was to change the way that radio worked completely. Instead of being a two-way medium in which operators chatted with each other, it swiftly evolved in the postwar period into a centralized, one-way, broadcast medium. Gernsback's article "Amateur Radio Restored" concluded with the words "Long live the radio amateur!" But as he wrote these words the golden age of open, two-way wireless was in fact coming to an end.

## THE ORIGINAL WWW

Now that continuous-wave transmitters could transmit audio, it was no longer necessary to understand Morse code to use a radio. The fun of scanning the airwaves with a radio receiver was suddenly open to all. Broadcasting—sending out a general signal for anyone to tune into—started to become popular. Concerts and speeches were broadcast experimentally on both sides of the Atlantic, starting in 1919. In June 1920 a concert in England by the soprano Dame Nellie Melba was transmitted from the Marconi station in Chelmsford, Essex, and was picked up by "amateur wireless telephone operators throughout Europe," according to *Telephony* magazine. "There is, of course, a great deal to be done before such methods can be established on a commercial scale," declared the *Electrician*, "but there is no reason why any time should be lost in educating the public in the progress of this new means of inter-communication." In fact, commercial broadcasting took off more quickly than anyone expected. Initially, American amateur stations also tried their hand, but in December 1921 frequencies for broadcast entertainment were established to keep them clear of naval distress frequencies, and the next month amateurs were "temporarily" banned from entertainment broadcasting. The ban was in fact permanent, and during 1922 radio underwent a radical transformation.

This shift had its origins during the war, when U.S. government officials began to try to establish a government monopoly in radio. They were particularly worried about the dominance of the Marconi Wireless Telegraph Company, a British firm. Britain dominated the world's wired telegraphy industry and now a British firm was, it seemed, about to extend that dominance to wireless. The war meant that all radio stations in America had been placed under the control of the U.S. Navy. To complete its dominance of the radio industry the navy also quietly bought the wireless stations operated by the Federal Telegraph Company, a private company based in California, without asking for the approval of Congress. Lawmakers were unimpressed by the navy's attempted takeover of the radio industry and ordered it to reverse the transaction. They then established a new national champion by forcing the American arm of Marconi to sell its operations to General Electric, and then spinning off GE's radio assets into a new company, the Radio Corporation of America. RCA was also granted the right to use more than two thousand radio-related patents belonging to other firms including Westinghouse, a maker of electrical equipment, and American Telephone & Telegraph (AT&T), the company that had come to dominate America's telegraph and telephone systems. RCA's aim was to establish a wireless-communications service between America and the rest of the world, and it took as its logo a globe with the words "World Wide Wireless" superimposed. At the time the cost of sending a telegram to England was twenty-five cents a word. With its high-power radio equipment and agreements with wireless telegraphy firms in other countries, RCA was able to undercut wired-telegraph rates by about 30 percent.

But RCA's focus switched from Morse-code telegraphy to broadcast entertainment in a matter of months. After the first experimental broadcasts in 1919 and 1920, radio stations were established in many cities, driving demand for receivers to pick up their broadcasts. Overall sales of radio receivers in America in 1922 were sixty million dollars, corresponding to around one million build-your-own receivers and one hundred thousand off-the-shelf radios. Radio sales increased to $136 million in 1923, $358 million in 1924, and exceeded half a billion dollars in 1926. By this time off-the-shelf radios were outselling kit radios. By 1928 there were more than thirteen million receivers in use. This sudden liftoff was known as the "Euphoria of 1922." Newspapers breathlessly reported the launch of each new broadcast station and every technological breakthrough associated with radio. Books were rushed out to help people understand and enjoy the new medium, such as *Radio for Everybody* and *Radio Receiving for Beginners*, which declared:

> Radio is a most fascinating pastime. While its possibilities are so boundless and intricate that the most advanced engineers are only beginning to solve them, you may begin with very simple operations which may be explained in simple terms. Thousands of twelve year old boys, and girls, are operating satisfactorily, entertaining their families and friends, while educating themselves. Anyone with common sense can readily grasp the elementary principles and begin receiving at once . . . One has a constant sense of participating in the performance of magic. Nobody starts receiving and abandons it. One step leads to the next, and you simply cannot let it alone. All the while you are playing with one of the most worth-while developments of the age.

Speculation abounded over the long-term impact of this new medium. As RCA's annual report in 1922 noted, when the company had been set up, "it was not at that time foreseen that the broadcasting art would ever reach the high point of popularity that it has in the last year. The engineers and scientists had anticipated the development of wireless telephony for communication purposes, but no one had visualized the phenomenal expansion of wireless telephony as used today for broadcasting." But during 1922 the number of broadcasting stations in the United States had grown from fewer than twenty to nearly six hundred. "The art itself is advancing very fast," the report observed, "and the ultimate effect of broadcasting upon the economic, social, religious, political, educational life of the country and the world, is comparable only with that of the discovery of printing 500 years ago." Of the 576 broadcast stations, some were operated by amateurs; others were backed by makers of radio receivers such as RCA, AT&T, and Westinghouse; by government agencies; or by universities. The number of amateurs licensed for two-way communication was still far higher, at 16,570. But 1923 proved to be the peak for the number of amateur stations, which subsequently went into decline. Receivers already far outnumbered transmitters. Within a very short time radio had become a predominantly one-way, broadcast medium.

Despite the rapid take-up of the new technology, it was initially unclear exactly how anyone would make money from it. Makers of radio receivers launched stations to ensure a market for their products; universities and government agencies launched stations as public-service ventures; and department stores, newspapers, and car dealers launched them as a form of self-promotion. As early as 1922, debate raged over how the medium could be made financially self-sustaining. "Free broadcasting services obviously cannot go on forever," noted *Popular Science Monthly* in June 1922. Some people proposed various forms of radio subscriptions; others advocated a tax on every radio receiver sold, or selling licenses to allow people to use receivers.

Another proposal was to support the new medium through advertising. Initially this idea was deeply unpopular. Those expressing their distaste for advertising included industry journals, government officials, ordinary members of the public, and the chairman of RCA, David Sarnoff. The commerce secretary, Herbert Hoover, declared in 1922 that "it is inconceivable that we should allow so great a possibility for service to be drowned in advertising chatter." A spokesman for Westinghouse said advertising "would ruin the radio business, for nobody would stand for it." It was not even clear whether radio advertising was legal, because AT&T claimed that the sale of airtime to advertisers was a kind of paid-for telephone service, over which it held a monopoly. In 1924 AT&T linked up sixteen of its radio stations using its long-distance telephone network, enabling it to create a national network whose signals could reach 65 percent of the U.S. population (provided, that is, that they owned a radio). Aggregating the audiences of multiple radio stations in this way cut the cost of producing radio programming, while also allowing AT&T to charge much higher advertising rates. The bigger a radio network was, the more money it could make. As the owner of the telephone network, however, AT&T could prevent RCA and other broadcasters from following suit.

Sarnoff launched a legal challenge against AT&T, pointing out that it had handed over its radio-related patents when RCA was established in return for a minority stake in the company, and did not have the legal right to make or sell radio receivers. The arbitrator overseeing the case agreed with RCA, and the ruling came just as antitrust regulators launched an investigation into the radio industry. Outmaneuvered by Sarnoff and worried that its attempt to move into radio might threaten its telephone monopoly, AT&T capitulated, selling its radio stations to RCA, agreeing to provide the long-distance lines needed to link them up, and abandoning its claim to a monopoly on radio advertising. The result was the first genuinely national radio network, the National Broadcasting Company (NBC). With its national audience it could appeal to advertisers and could afford to hire Broadway stars and other popular artists to perform on air. At the same time, local stations that were part of the network could sell local advertising.

Radio advertising was eagerly embraced by a cash-strapped industry, and the model of private stations supported by advertisers came to be known as the "American plan." The Federal Radio Commission (FRC), established in 1927 to administer America's airwaves, encouraged further centralization by favoring large broadcast networks over smaller local stations. It reorganized the radio band to create forty nationwide channels, shunting smaller stations off onto other parts of the dial. The nationwide broadcasters organized themselves into a handful of national networks, notably those operated by NBC and its main rival, the Columbia Broadcasting System (CBS). The FRC said that because of the limited amount of radio spectrum, it favored general-interest commercial radio stations over specialized noncommercial ones, arguing that "There is not room in the broadcast band for every school of thought, religious, political, social, and economic, each to have its separate broadcasting station, its mouthpiece in the ether."

Radio centralization occurred in other countries, though for a different reason. In Britain broadcasting was a monopoly from the start, with the foundation in 1922 of the British Broadcasting Company, the world's first national broadcaster. It was formed when the General Post Office, along with Marconi, General Electric, and other companies, decided to pool their assets into a consortium. Daily broadcasts began that year using a transmitter installed on the roof of Selfridges, a department store on Oxford Street in London. The BBC's first general manager, John Reith, explained that Britain had learned lessons from America when setting up its broadcasting system. "America showed us what pitfalls to avoid; we learnt from experience," he observed. "The lack of control in America was resulting in chaotic confusion . . . Britain, as I say, benefitted by America's example, and a centrally controlled system of broadcasting stations was the result." Rejecting the American advertising-funded approach, a government committee in 1923 decided that a license fee of ten shillings was the best way to fund the BBC. In 1927 the BBC was reestablished under a royal charter as a state broadcaster, operating at arm's length from the government and renamed the British Broadcasting Corporation.

Together with its royal endorsement, its own coat of arms, and its status as an arm of the government, the BBC had a paternalistic and condescending attitude toward its audience. It wanted to improve and educate its listeners, not merely entertain them. Like the diners at a genteel dinner party, the eminent thinkers who were invited to give lectures on air were encouraged to avoid controversial topics such as politics and religion. The pronunciation used on air was expected to be exemplary and was overseen by an Advisory Committee on Spoken English of experts including academic linguists and famous writers. In 1931, for example, the committee recommended that the pronunciation of the word *garage* should rhyme with "marriage" rather than "barrage." Reith thought the accent and pronunciation of the aristocracy would be most easily understood by listeners at home and abroad, but it added to the perception that the BBC was snooty and elitist. The BBC considered its aim to be to lead public taste, not to follow it; to give people what it felt they ought to want, rather than what they wanted. It came to see itself as the guardian of cultural standards, keeping them free of commercial or popular pressures. In his book *Broadcast over Britain*, published in 1924, Reith explained his position.

> As we conceive it, our responsibility is to carry into the greatest possible number of homes everything that is best in every department of human knowledge, endeavour, and achievement, and to avoid the things which are, or may be, hurtful. It is occasionally indicated to us that we are apparently setting out to give the public what we think they need—and not what they want, but few know what they want, and very few what they need.

The establishment of the BBC served as a model for many other European countries, which also set up state-run monopoly broadcasters funded by license fees. By 1936, twenty-two countries (mostly in the Americas) had adopted the American approach of advertising support; but more than forty countries, mostly in Europe and Asia, had embraced the "BBC system" of government-funded radio, supported by a tax on radio receivers. This approach was viewed with mistrust in America, because it centralized control of broadcasting in government hands. But in many countries that was exactly what governments wanted.

The trend toward centralized government control of radio was most pronounced in Germany. In 1932, in the dying days of the Weimar Republic, the various semicommercial regional stations that had emerged in the 1920s were nationalized and combined to form the Reich Broadcasting Corporation. Advertising was banned and political content was favored. This played to the hand of the rising Nazi party and its charismatic leader, Adolf Hitler, whose rousing oratorical style was a perfect match for radio broadcasting. When the Nazis came to power in 1933 the RBC became central to their propaganda operation. Joseph Goebbels, Hitler's head of propaganda, outlined his views on radio in a speech in August at the opening of a radio exhibition. The new government had subsidized the development and production of a low-cost radio receiver, the Volksempfänger (people's receiver), so that even the poorest households could receive its political broadcasts. Millions of these radios were sold. They were designed so that they could pick up German domestic broadcasts easily but were unable to receive foreign broadcasts on shortwave bands. Goebbels promised "a systematic campaign" to promote the new radios. "We will use the knowledge of propaganda we gained in the past years. Our goal is to double German radio listenership," he said in his speech. Goebbels was explicit about radio's potential to effect political and societal change: "The radio will be for the twentieth century what the press was for the nineteenth century," he declared. Under the Weimar Republic the government had been "blind to the possibilities of this modern method of influencing the masses . . . We on the other hand intend a principled transformation in the worldview of our entire society, a revolution of the greatest possible extent that will leave nothing out, changing the life of our nation in every regard . . . It would not have been possible for us to

A Nazi-era poster advertising the *Volksempfänger*, or "people's radio," under the slogan "The whole of Germany hears the Fürher with the People's Radio." *Mary Evans Picture Library/Weimar Archive*

take power or to use it in the ways we have without the radio and the airplane."

Goebbels complained that under the Weimar Republic the administration of radio had become overly bureaucratic. The Nazis, he said, would "eliminate excessive organization as quickly as possible, replacing it with Spartan simplicity and economy" and putting the new medium to work in the interests of the nation.

> We do not intend to use the radio only for our partisan purposes. We want room for entertainment, popular arts, games, jokes, and music. But everything should have a relationship to our day. Everything should include the theme of our great reconstructive work, or at least not stand in its way. Above all it is necessary to clearly centralize all radio activities . . . We want a radio that reaches the people, a radio that works for the people, a radio that is an intermediary between the government and the nation.

Radios were placed in offices, factories, and public places, making the endless coverage of speeches and rallies almost inescapable. Writing in *Foreign Affairs* in 1938, the American journalist César Saerchinger described the Nazis' use of radio. "In their hands it has become the most powerful political weapon the world has ever seen. Used with superlative showmanship, with complete intolerance of opposition, with ruthless disregard for truth, and inspired by a fervent belief that every act and thought must be made subservient to the national purpose, it suffuses all forms of political, social, cultural and educational activity in the land." When the Second World War began, the Nazis made listening to foreign broadcasts a crime. Only the intelligence services were allowed to listen to foreign stations. But many Germans did so to find out what was really going on; the heavily censored official news channels gave no inkling that Germany was losing the war. As late as April 1945, Goebbels's broadcasts told listeners that a great victory was within reach. The truth only became apparent to most Germans with the announcement, via radio, of Hitler's death. After the war Albert Speer, one of Hitler's ministers, told the Nuremberg trials: "Hitler's dictatorship differed in one fundamental point from all its predecessors in history. His was the first dictatorship . . . which made the complete use of all technical means for domination of its own country. Through technical devices like the radio and loudspeaker, eighty million people were deprived of independent thought. It was thereby possible to subject them to the will of one man."

In the United States radio was centralized to maximize advertising revenue; in Britain to preserve and promote the values of the elite; and in Germany to advance Nazi propaganda. Whatever the reason, the result was the most centralized medium in history. In the United States radio listeners were gathered up by networks that saw them as consumers to be sold to; in Britain they were the masses to be instructed and improved; in Germany they were the people to be indoctrinated and misled. In each case there was a striking "us and them" division between broadcasters and the faceless mass of their listeners.

## TELEVISION, THE DRUG OF THE NATION

The idea of sending pictures, in addition to sound, from one place to another in real time was widely assumed to be the next step beyond telephony as early as the 1880s—though it was assumed, at the time, that these pictures would travel along wires. By the time the first practical demonstrations of television took place in 1926, by John Logie Baird in Britain and Charles Francis Jenkins in America, the broadcast model of radio was well established. Although some inventors experimented with two-way television, it never had a free-for-all period in the way radio did, both because radio transmissions had become tightly regulated by the 1920s, and because the technology needed to encode and transmit images was so much more elaborate. Given its expense and complexity is it not surprising that television was expected to be a centralized, broadcast medium from the beginning. But it did not merely import content, conventions, and funding models from radio; it also ended up in the hands of the existing radio empires.

Predictions of the forthcoming wonders of television began to appear in the 1920s, in some cases before it had even been shown to work. Newspapers reported on progress toward "radio optics" or "sight radio." A book entitled *Broadcasting: Its New Day*, published in 1925, predicted that "the ether will vibrate with the likenesses of our favorite stars, which we will receive faithfully . . . Our baseball players, instead of performing before a group of spectators, will perform before a radio transmitter and we shall hear the whack of the bat and the call of the umpire, and see the dust raised by the sliding player's feet. Radio vision is not an idle dream." One pundit declared in 1925 that all American households would have televisions by the end of the decade. And this was before anyone had built a working example of the technology.

The first working television system was built in Britain, where Baird demonstrated his prototype to a delegation of scientists in his laboratory in January 1926. His apparatus, which built on the work of many previous inventors, used a spinning disk, perforated with a spiral of carefully placed lenses, to scan the target scene in succession. The resulting variations in brightness, measured using a photocell, were transmitted to a receiving station. The image was then reconstructed by varying the brightness of a lamp viewed through a second spinning disk, synchronized to the first. Other inventors had successfully transmitted crude images and silhouettes, but Baird's system was the first capable of depicting motion in real time. A reporter for the London *Times*, who attended the 1926 demonstration, wrote that the resulting image was "faint and often blurred, but substantiated a claim . . . it is possible to transmit and reproduce instantly the details of movement, and such things as the play of expression on the face." Soon afterward Jenkins staged a public demonstration of his "radiovisor," which was also based on a spinning disk, in Washington, D.C. He also transmitted moving images of a toy windmill over a short distance by radio.

Both men set about trying to commercialize their inventions. Jenkins established the first television station, W3XK, in Washington, D.C., in 1928. Mindful of the power of the network model used by radio, he opened a second station in New Jersey the following year. He also began selling television receivers, both ready-made and in kit form, in an effort to build up an audience, following the example set by the pioneers of radio. An advertisement for Jenkins's receivers invoked memories of the golden age of experimental radio: "Television is here! It is ready for the experimenter, service man, and dealer! Television programs are steadily improving. Now is the time to get into television. Experience the thrills of pioneer broadcast days all over again!" Jenkins's work prompted General Electric and AT&T, which were developing their own versions of the technology, to launch experimental television stations, too.

Baird, meanwhile, demonstrated long-distance transmission, using telephone lines to send images over four hundred miles from London to Glasgow and shortwave radio to broadcast across the Atlantic. He participated in television experiments in France and Germany and set up his own company, Baird Television, which was granted permission by the BBC to make experimental broadcasts using its powerful transmitter. But television required two transmitters (a separate one was needed for the sound), so to start with these transmissions consisted of images and sound alternating every few minutes. Simultaneous sound and vision became possible in 1930 when the BBC added a second transmitter. In 1932 the BBC took over responsibility for creating programs, and by 1935 there were several thousand enthusiasts tuning in to its broadcasts.

In America, however, Jenkins had run into a powerful adversary: David Sarnoff of RCA. Sarnoff could see that television had the potential to overshadow radio, and he wanted to ensure that RCA and its national network, NBC, maintained their dominant position. He and other broadcasters regarded television as a logical extension of radio and did not want newcomers venturing onto their patch. RCA argued that "only an experienced and responsible organization such as Radio Corporation of America, should be granted licenses to broadcast material, for only such organizations can be depended on to uphold high ideals of service." Sarnoff also observed, with some justification, that the electromechanical television systems being tested around the world were crude and not yet suitable for mass adoption. A combination of RCA's lobbying, immature technology, and the onset of the Depression killed off efforts by Jenkins and others to get television off the ground in America in the early 1930s.

For many years inventors had been working on a superior technology that did not rely on rapidly rotating disks, or indeed on any mechanical moving parts at all. Instead they hoped to capture and display images using electron beams in cathode-ray tubes. The first to get this all-electronic approach to television to work was Philo T. Farnsworth, a self-taught inventor whose interest in the subject began when he found a pile of popular-science and radio magazines in the attic of his family's new home on a farm in Idaho. He later claimed that it was when he was plowing a potato field, at the age of fifteen, that he first had the idea that images could be broken down and reconstructed by scanning them line by line, like a plow moving across a field. Six years later he had built a working television system, based around an image sensor he called an "image dissector," and he demonstrated it to the press in his laboratory in San Francisco in September 1928. The *San Francisco Chronicle* described the image dissector as being the size of "an ordinary quart jar that a housewife uses to preserve fruit." It was much smaller and simpler than the rotating disks championed by Baird and Jenkins. The only drawback of Farnsworth's system was that it required the use of extremely powerful illumination.

The publicity around Farnsworth's work brought it to Sarnoff's attention at RCA. Sarnoff had just hired Vladimir Zworykin, a Russian inventor who had been working on electronic television at Westinghouse for many years. Zworykin spent three days visiting Farnsworth's laboratory, pretending that he was interested in investing in his invention, and learning all he could. With funding from RCA, Zworykin then spent several years replicating and ultimately improving upon Farnsworth's invention. In particular, he built a more sensitive image detector that did not require the use of such powerful lights. Meanwhile, Sarnoff immobilized Farnsworth with a series of lawsuits, claiming that Farnsworth's patents were invalidated by Zworykin's earlier work at Westinghouse. In 1939 Farnsworth finally prevailed in a decade-long legal battle, and RCA was ordered to pay one million dollars to license his invention. But it was a Pyrrhic victory for Farnsworth, for RCA had stifled the adoption of his technology and was by this time preparing to launch its own television system, unveiled with great fanfare at the New York World's Fair earlier that year. At a press conference held in RCA's pavilion at the fair, Sarnoff announced "the birth in this country of a new art so important in its implications that it is bound to affect all society." He ended his speech: "Now, ladies and gentlemen, we add sight to sound!"

The curtains on the stage were thrown back to reveal rows of television screens, all showing a live image of Sarnoff. It was a theatrical masterstroke, demonstrating at once that television was no longer an experimental technology and planting in the public consciousness RCA's claim that television was radio with pictures, and was therefore the natural province of radio broadcasters. At last it seemed that the grand promises about the coming marvels of TV, which had continued throughout the economic gloom of the 1930s, would be realized. In a promotional film shown at the World's Fair, RCA declared the technology ready for mass adoption: "The research problem of yesterday is the radio marvel of today. Another milestone of progress has been passed, and science has made a reality of the age-old dream of pictures from the sky." In reality, however, the launch of TV was a flop. RCA hoped to sell one hundred thousand TV sets by the end of 1939 to establish its technology as the standard; in the event it sold only three thousand. The first sets were expensive, some costing as much as a small car, and signals could only be picked up close to a small number of receivers.

Then the Second World War intervened. Electronics manufacturers focused on making military equipment rather than devising new consumer products. A handful of experimental stations stayed on the air in America during the war, albeit with reduced hours, but the BBC shut down its television operation completely during the conflict, so that German bombers heading toward London could not use its TV transmitters as navigational beacons. When transmissions resumed after the war, the new medium was firmly in the hands of established radio broadcasters: NBC and CBS with their familiar advertising-supported model in America, and the state-broadcaster model, with funding from a tax on television sets, in Europe. From an organizational perspective, television really was just radio with pictures.

Television took off mostly rapidly in postwar America, where its success had come to be seen as inevitable. During the war, an advertising blitz convinced Americans that this magical new technology was a reward that awaited them once the conflict was over. Makers of electrical appliances depicted the postwar future as a time when every home would be equipped with labor-saving and entertainment products. RCA's wartime advertising promised that television was just around the corner: "America's 'Next Great Industry' awaits only the green light of Victory to open up undreamed-of horizons in Education, Entertainment, Employment," it maintained. RCA's magazine advertisements in 1944 proclaimed that television was "the Baby that will start with the step of a Giant." Those for DuMont, another electronics manufacturer, declared in 1944 that "but for the war, a DuMont Television-Radio Receiver might have been your most exciting gift this Christmas! We shall do our best, God willing, to bring you and yours the gift of fine quality television before the next Christmas Season." NBC's ads called television "A season pass to baseball" and asked readers to "Imagine Bob Hope on Television." A poll at the end of the war showed that 85 percent of Americans knew what television was, though only 19 percent had ever seen it. Even government officials talked up television: in 1942 James Fly, head of the Federal Communications Commission, the body that now regulated American broadcasting, suggested that "during the post-war period television will be one of the first industries arising to serve as a cushion against unemployment and depression . . . There is no reason now apparent why we should not aim at a 50-million set television industry, mirroring the present 50-million set standard broadcast industry."

With its heavy marketing of television, even before it was available, RCA wanted to do more than simply ensure there would be pent-up demand for the technology once the war was over. It also wanted to ensure that its own formats and technologies, devised before the war, would become the standard. Its rival CBS was lobbying for a slower rollout of television, to give it longer to catch up and develop its own technology. It was behind RCA but hoped to leapfrog it by developing color TV. CBS also advocated the use of ultra-high-frequency (UHF) spectrum, which would allow for more channels and higher-quality transmissions. The FCC backed RCA, however, and its proposal to go ahead with television in its black-and-white form, on existing very-high-frequency (VHF) bands. (The appearance of coziness between RCA and the FCC was underlined a few months later when the chairman of the FCC resigned and went to work at NBC.) The use of VHF rather than UHF spectrum ensured that television would be a national medium with a small number of channels rather than a local medium with a larger number, which was exactly what RCA and NBC wanted. Independent TV stations could survive only in large cities where there were lots of viewers whom advertisers would pay to reach. Most television would be network television, with programming devised to have the broadest possible appeal, rather than catering to specialist interests. In the first three decades of television, the networks accounted for 70 to 90 percent of viewing time in America.

Television really took off in 1948–49, when the price of a set fell to the equivalent of six weeks' wages. The number of televisions manufactured in the United States went from 6,500 in 1946 to 179,000 in 1947, 975,000 in 1948, and 1.7 million in 1949. During 1948 the advertising model also solidified: initially advertisers were asked to sponsor

An American advertisement for television that appeared during the Second World War. *Courtesy of J. Fred MacDonald*

programs by paying their production costs, but in 1948 there was enough demand that television stations could charge advertisers for television time. Advertisers favored national shows aired on networks, however, and by the end of 1949 networks were taking half the advertising revenue. Local television was unsustainable, and by 1953 nearly all TV stations in America were affiliated with networks. NBC, CBS, and ABC (the American Broadcasting Company, a new network spun out of NBC in 1943 at the behest of antitrust regulators) staffed their television operations with technicians and talent from radio. Many television shows were simply transplants from radio. In some cases scripts for soap operas and dramas were recycled word for word; in other cases radio and TV versions of the same show were broadcast simultaneously.

The proportion of American households with televisions jumped from 9 percent in 1950 to 65 percent in 1955. Color broadcasts began in 1954, the subject of another bruising fight between RCA and CBS, in which David Sarnoff once again emerged victorious. Household penetration of television sets reached 87 percent in 1960 and 95 percent in 1970. At this time only around half of all sets were color sets, but about one third of America homes had more than one set. No new invention had ever entered American homes more quickly. Radio took six years to go from 10 percent to 40 percent penetration, from early adopters to mainstream acceptance; television took just three. In other countries the takeoff came a little later. In Britain, where the BBC adopted a new standard very similar to that used by RCA, television's fortunes were boosted by the coronation of Queen Elizabeth II in 1953, the first coronation ceremony ever televised. By 1960 there were three televisions for every ten people in the United States, and two for every ten people in Canada and the United Kingdom. Television was also taking off in Western Europe, Japan, and Australia.

But the United States was the unquestioned leader in the development, adoption, and usage of television. American programs and formats, from detective dramas to quiz shows, became popular around the world. The networks' desire to maximize their audience ratings, and hence their advertising revenue, led critics to argue that the early promise of television, as a medium that would promote education and strengthen democracy, had been thrown away. The attack on television by Nicholas Johnson, a former official at the FCC, was particularly noteworthy because it came from an industry insider. In his book *Test Pattern for Living*, published in 1972, he declared television an unmitigated failure.

> Television was seen as an opportunity to extend the perception of the average American, to open for all the great excitement and education and self-fulfilling potential that can come from exposure to the best that man has to offer. By now almost everyone, including those in the industry, would concede that television has failed. Not only has it failed to make of us a better race of men, it has actually made us worse than we were before. The former would be indictment enough. The latter is simply intolerable.

Johnson's critique of the "pervasive and invidious . . . anesthetizing effect" of television was echoed by other critics. Television was blamed for causing crime and social disharmony. In *Amusing Ourselves to Death*, published in 1985, the American media theorist Neil Postman likened the situation to that depicted in Aldous Huxley's novel *Brave New World*, in which the people were oppressed not by an authoritarian government, but by their addiction to entertainment. Like Johnson, he complained that even television news was really a form of entertainment, the real aim of which was to keep people watching the lucrative advertising breaks. Television's appeal is apparent from the steady increase in the average amount of time spent watching television in America, from four and a half hours a day in 1950 to five hours in 1960, six hours in 1970, and seven hours in 1990. As the number of homes with multiple screens increased, and cable and satellite television provided dozens and then hundreds of channels to choose from, the number of hours watched increased still further, exceeding eight hours a day in the early twenty-first century.

Television had become the most pervasive medium ever. The couch potato, vegetating in front of the flickering screen, emerged as a cultural cliché. Watching television, an entirely one-way, passive experience, became the very definition of inaction. Only sleeping involves less effort. The broadcast model considers the role of the radio listener and television viewer to be merely that of a passive consumer. This is as far as it is possible to be from a media system in which people create, distribute, share, and rework information and exchange it with each other. It is the opposite of social media.

# CHAPTER 11

## THE REBIRTH OF SOCIAL MEDIA: FROM ARPANET TO FACEBOOK

*The Internet is proving to be one of the most powerful amplifiers of speech ever invented. It offers a global megaphone for voices that might otherwise be heard only feebly, if at all. It invites and facilitates multiple points of view and dialogue in ways unimplementable by the traditional, one-way, mass media.*

—Vinton Cerf, 2002

### FROM ARPANET TO THE INTERNET

IN RETROSPECT, THE inauguration of what would turn out to be the largest and most complex communications system in human history could have gone a little more smoothly. But at the time none of the people involved were aware of the significance of what they were doing. It was the evening of October 29, 1969, and Charley Kline, a twenty-one-year-old student at the University of California, Los Angeles, was working late in the computer lab. UCLA had an SDS Sigma 7 computer, a mainframe that filled an entire room. Several people sitting at separate terminals could use this giant computer at the same time, and Kline could be found writing code on it at all hours of the day and night. That evening Leonard Kleinrock, the professor in charge of the computer lab, asked Kline to help him test a new device that would link the Sigma 7 to another computer at the Stanford Research Institute, four hundred miles away in Menlo Park, California.

The project to link computers in this way had begun when Bob Taylor, an official at the Advanced Research Projects Agency (ARPA), the research arm of the U.S. Department of Defense, became frustrated by the proliferation of computer terminals in his office. ARPA was funding computer projects at the University of California, Berkeley; at the Massachusetts Institute of Technology (MIT); and at the System Development Corporation, a pioneering software company based in Santa Monica. In his office Taylor had separate terminals connected by fixed telephone lines, called leased lines, to each of their computers, so that he could monitor the work being done. As ARPA continued to fund more computer-related projects, it seemed that Taylor's office would soon be overrun with terminals. Worse, because his three terminals were connected to different computers, he could not transfer information from one to another except by laboriously retyping it. "I said, oh, man, it's obvious what to do: If you have these three terminals, there ought to be one terminal that goes anywhere you want to go," he later recalled.

Being able to access remote computers using a single terminal would also, Taylor realized, make it easier for researchers to communicate with each other. As soon as the first multiuser computer systems had appeared in the early 1960s, people had started using them to send messages to each other. With three terminals in his office, Taylor was well aware of the power of such "electronic mail" to encourage collaboration and the exchange of information. As he told the *New York Times* in 1999,

> The thing that really struck me about this evolution was how these three systems caused communities to get built. People who didn't know one another previously would now find themselves using the same system. Because the systems allowed you to share files, you could find that so-and-so was interested in such-and-such and he had some data about it. You could contact him by e-mail and, lo and behold, you would have a whole new relationship.

Early e-mail systems only allowed the exchange of messages between people using the same mainframe computer, however. Although a community grew up around each mainframe, there was no way to send a message to someone on another machine. Taylor proposed that ARPA fund the construction of an experimental computer network to link machines at different sites. If it worked, this new network, to be known as ARPANET, would not simply allow him to reclaim some space in his office. It would encourage communication and cooperation between researchers at different institutions, letting them exchange information and making their powerful computers more widely available to other researchers. The aim of the ARPANET, in short, was to connect people as well as machines.

Taylor drew up a detailed plan for the network, proposing that it use a promising new theoretical approach called "packet switching." Instead of connecting all the machines on the network directly to each other with leased lines, this involved breaking data down into small, uniform "packets" that could be passed from one machine to another until they reached the appropriate destination. A computer could then talk to a distant machine via several intermediaries, without needing a direct connection. This approach would greatly reduce the number of leased lines needed to interconnect a given number of computers, and would also make efficient use of network capacity by interleaving traffic between multiple sources and destinations. Packet switching was originally proposed as a way to build networks that would keep working in the event of a nuclear attack, because when part of a packet-switching network is disabled, packets can simply be routed around the problem, finding another path to their destination. But given the cost and unreliability of network links and computer hardware in the 1960s, computer scientists realized that packet switching was also a good way to build reliable networks for general use.

Taylor invited one hundred and forty companies to bid for the contract to build special interface boxes, called "Interface Message Processors" (IMPs), which would be plugged into computers at different sites and linked up by leased lines. Industry giants IBM, the biggest provider of mainframes, and AT&T, America's telecoms monopoly, declined to bid. Rather than interconnecting and sharing separate computers, IBM imagined a future of ever-larger mainframes, with remote terminals connected by AT&T's lines, as the best way to bring many users together. The contract for the IMPs was ultimately awarded to a small company called Bolt, Beranek and Newman. By the middle of 1969 it had built four IMPs, each the size of a large cupboard, and had delivered them to four test sites, including UCLA and the Stanford Research Institute. Kleinrock, who had done a lot of work on the theory of packet switching, was responsible for connecting an IMP to UCLA's computer and then testing it by establishing a link to Stanford. He asked Kline to help.

"We should have prepared a wonderful message," Kleinrock said many years later. "Certainly Samuel Morse did, when he prepared 'What hath God wrought,' a beautiful Biblical quotation . . . or Armstrong up in the moon: 'a giant leap for mankind.' These guys were smart. They understood public relations. They had quotes ready for history." The message that would be sent along the link from UCLA to Stanford, however, was rather more mundane: it was simply the word "login," which would enable Kline to log into the Stanford computer from his terminal at the UCLA computer. At about 9:30 P.M. Kleinrock called Bill Duvall, a young programmer at Stanford, to say that they were ready to begin the test. Kline typed an "L" on his terminal. Over the phone, Duvall confirmed that it had successfully traveled through the two IMPs to reach the Stanford computer. Kline then typed an "O." Again, Duvall confirmed that it had been received. Kline typed a "G." But nothing happened. The system had crashed. The first message sent across the ARPANET was therefore "LO." After some fiddling, they tried again, and at 10:30 P.M. they successfully established a link that enabled Kline to log into the Stanford computer. The IMPs worked as intended, and the ARPANET was born.

The IMPs at UCLA and Stanford were permanently connected on November 21. By early December they had both been linked to a third IMP at the University of California, Santa Barbara, and the Stanford IMP had been connected to the fourth IMP at the University of Utah. The packet-switching system meant that users at each of the four sites could access any of the four connected computers, even when there was not a direct link between their respective IMPs. (Network traffic between UCLA and the University of Utah, for example, traveled via Stanford or Santa Barbara.) ARPANET was extended to the east coast in March 1970 and continued to grow as more computers, connected by more IMPs, were added to the network. In 1975, when ARPANET was declared fully operational, rather than being an experimental project, there were 57 IMPs, including one across the Atlantic in London. By 1981 there were 213 computers attached to the network, with another being added, on average, every twenty days. In January 1983 the packet-switching protocol used by the IMPs, known as NCP, was retired in favor of a more robust standard called TCP/IP, which had been developed by Robert E. Kahn and Vinton Cerf in the early 1970s. It provided a common "internetwork protocol" to smooth over the differences between the ARPANET and

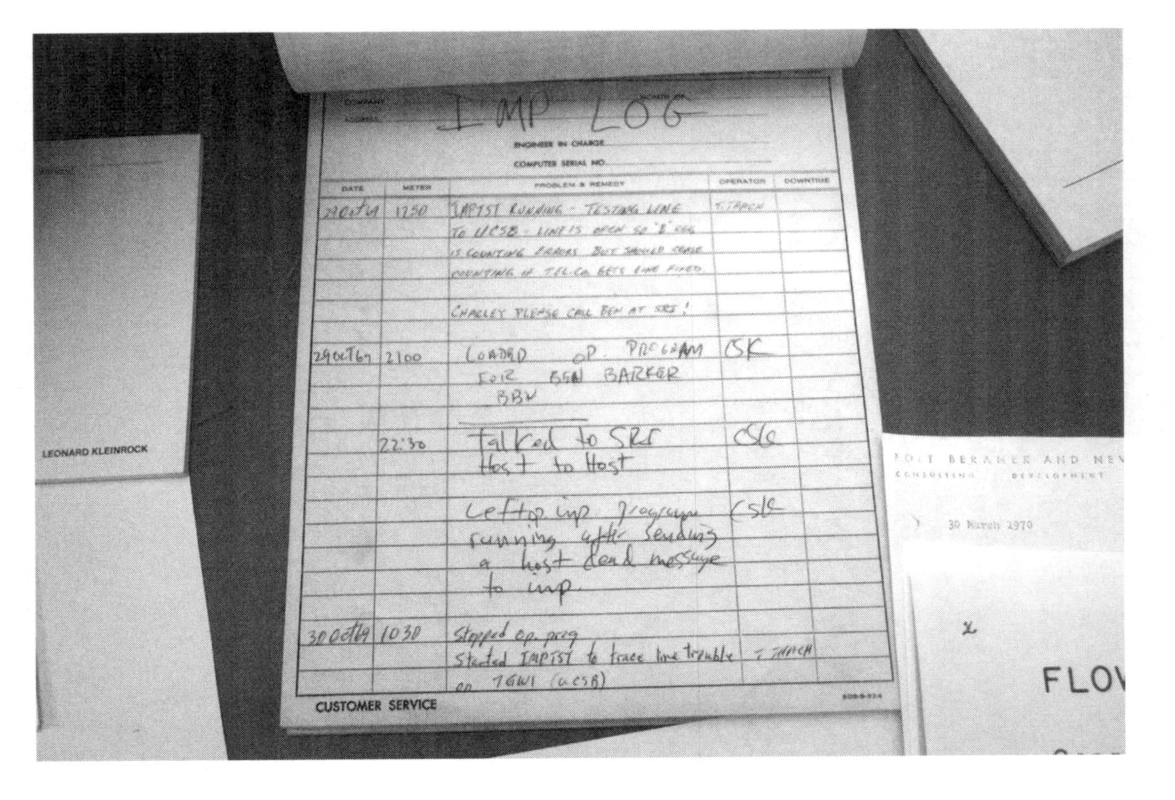


Log from the UCLA Interface Message Processor recording the sending of the first message across ARPANET. *Courtesy of FastLizard4 on Flickr, licensed under the CC-BY-SA 2.0 license*

other packet-switched networks that had, by this time, emerged alongside it. The practice of linking entire networks, rather than individual computers, came to be known as "internetworking" or "internetting." By the early 1980s the network of interlinked, packet-switched computer networks, including the original ARPANET, had become known collectively as the Internet.

Once ARPANET had been established, users could log into remote machines and maintain e-mail correspondence with colleagues at remote sites. The next logical step was to build an e-mail system that could deliver messages from one site to another, rather than requiring people to log into multiple e-mail accounts on different remote machines. The earliest such "network e-mail" system was developed in 1971, and the first messages were sent by Ray Tomlinson, a programmer at Bolt, Beranek and Newman, the company that built the IMPs. Tomlinson sent a few test messages between two machines that were in the same room, but were only connected via the ARPANET. Tomlinson has forgotten exactly what he wrote in the first message, but it was likely to have been some random keystrokes such as QWERTYUIOP. Although the message has been lost, one feature of Tomlinson's system survives to this day: his adoption of the @ symbol to separate the user name of the recipient and the name of the computer hosting the recipient's account. This "user@host" formulation has been used ever since.

As e-mail systems became standardized and interconnected, interpersonal messaging emerged as one of the most powerful and compelling features of the ARPANET, just as Taylor had anticipated. Very soon after Ray Tomlinson's invention of network e-mail, recalls Vinton Cerf, "people created mailing lists, one for science fiction and another one for restaurant reviews. And we could see instantly that e-mail was a social medium, in addition to simply being an interoffice memo system." Mailing lists allowed groups of people with similar interests to share information and engage in discussions using e-mail. Early mailing lists included science fiction lovers, wine tasters, and network hackers. Restaurant reviews, meanwhile, were compiled into a collaborative guide called YUMYUM, which was maintained by researchers at the Stanford Artificial Intelligence Laboratory. Any ARPANET user could send in a restaurant review for incorporation into the guide, which was made available for general download. Each reviewer was given a unique identifying number, which was added to the end of each of their reviews, so that readers could determine which reviewers they agreed or disagreed with, without knowing who they were. A version of YUMYUM from 1976 explains how this collective approach to authorship worked.

> This edition of YUMYUM lists 480 restaurants of the San Francisco Bay area and some outlying places together with 1,056 opinions from our readers. Many conflict with one another. Rather than attempting to form a "consensus," we let them tell it their way. Of course, different people are looking for different things and restaurants change with the passage of time and chefs . . . With a little experience, you can begin to evaluate the evaluators and draw more reliable conclusions from the opinions given here. We are dependent upon you to provide timely updates and information on interesting new restaurants. When preparing remarks for this guide, please try to express your views concisely and accurately . . . This guide is organized by geographical regions and types of cuisine. The coverage falls off approximately with the square of the distance from our laboratory.

The ARPANET's origin as a means of connecting ARPA-funded researchers at different universities meant that all communications were expected to be civil, like that between academic colleagues, and commercial or political use of the system was forbidden. A handbook written for newcomers to MIT's artificial intelligence laboratory in 1982 outlined the prevailing rules on "the DoD ARPANET/Internet," under the heading "Etiquette."

> It is considered illegal to use the ARPANET for anything which is not in direct support of Government business . . . personal messages to other ARPANET subscribers (for example, to arrange a get-together or check and say a friendly hello) are generally not considered harmful . . . Sending electronic mail over the ARPANET for commercial profit or political purposes is both anti-social and illegal. By sending such messages, you can offend many people, and it is possible to get MIT in serious trouble with the Government agencies which manage the ARPANET.

Universities that were not doing military research and were not connected to the ARPANET were unable to participate in this emerging online community, however. In 1979 this prompted two graduate students at Duke University, Jim Ellis and Tom Truscott, to propose a system that would allow distributed discussions to take place via regular file transfers between university computers, using temporary connections established over telephone lines, rather than the more expensive leased lines that linked ARPANET sites. The result was a decentralized discussion system called Netnews, later known as Usenet. The software that powered it was made freely available and spread quickly throughout the universities of North America. Much of the discussion on Usenet was technical in nature, but there were also discussions about a range of other subjects, subdivided into different discussion forums or "newsgroups." Any user could post a message into a group, or reply to an existing message. When browsing the contents of a given newsgroup, messages and their replies could be displayed together, to allow the thread of a discussion to be followed. Messages rippled across the system as computers on the network connected to pass on the latest updates, and Usenet was soon extended to ARPANET as well. The volume of Usenet traffic grew rapidly during the 1980s as the distinction faded between the military ARPANET and the wider Internet. And it was in newsgroups that flame wars (heated exchanges of messages), acronyms (such as LOL for "laugh out loud"), the custom of compiling lists of answers to "frequently asked questions" (FAQs), and the use of "emoticons" such as :-) first became widespread.

Like nineteenth-century telegraph operators, academic users of the Internet in the 1980s were a privileged minority with access to an open discussion environment that could be used without charge. But telegraphs never spread into the home. As personal computers proliferated in homes and offices in the 1980s, by contrast, their use as communications tools by the general public gradually became more commonplace. At first this involved dialing into simple bulletin-board systems that allowed participants to exchange messages; then it entailed joining online services such as CompuServe and America Online (AOL), which offered a range of information and services over dial-up connections; and finally, in the early 1990s, it became possible to sign up with a service provider to access the Internet. To begin with, direct Internet access from a personal computer was a fiddly business, requiring different software tools to manage connections, transfer files, handle e-mail, access newsgroups, and so forth. But access to the growing volume of information available online, along with the ability to commune with like-minded people, proved sufficiently compelling that some people, at least, were prepared to put up with the complexity of getting it all to work. And then, in 1993, it all suddenly got much easier, with the emergence into the mainstream of the World Wide Web.

## WEAVING THE WEB

When Tim Berners-Lee, a British scientist, wrote a program called WorldWideWeb in 1990, his aim was merely to make it easier for his fellow physicists to communicate with each other. Berners-Lee was working at CERN, the European particle-physics laboratory near Geneva, Switzerland, where a wide variety of computer systems were in use. In 1980 he had written a program called ENQUIRE, which let users jot down notes and ideas in the form of interlinked records called "cards." The card for a particular scientist might link to cards for scientific papers he had contributed to, for example, and the links also worked in the other direction, so that the card for a given scientific paper also linked back to the cards for its authors. Berners-Lee found this program useful in his own work, but few other people at CERN adopted it. Instead, the laboratory's thousands of researchers stored and exchanged information using several different and incompatible systems. Berners-Lee realized that collaboration would be much easier if there were a simple, universal way to interconnect them. In 1989 he began designing a new system based on the idea of hypertext, where individual words or phrases in a document can be linked to other documents, and in 1990 he wrote WorldWideWeb, a program capable of displaying and creating hypertext documents, called web pages, using new standards he had invented, such as HTML (hypertext mark-up language, which defined the formatting of a page) and HTTP (hypertext transfer protocol, which defined how the page should be requested by and delivered to the browser).

The very first web pages, created by Berners-Lee in 1990, did not contain images, just monochrome text with a choice of font sizes and styles. Crucially, however, text on one web page could be highlighted and turned into a clickable link to another page or document, either on the same computer or on a remote machine. His pioneering web browser could also display Usenet posts and navigate file repositories on remote servers, thus providing a unifying wrapper around a range of different computer systems used by researchers at CERN and elsewhere. This first browser had "back" and "forward" buttons and used the prefix http:// and the suffix .html in the addresses of web pages.

In early 1991 Berners-Lee made his WorldWideWeb browser and server software freely available over the Internet, at first to other researchers in the field of high-energy physics, and then, in the summer of 1991, to anyone else who was interested. Because he had written his software for the NeXTStep computer, which was not widely used, Berners-Lee and his colleague Robert Cailliau encouraged other people to write new versions that could run on different kinds of computers. One of those inspired by the possibilities of this emerging web of documents was Marc Andreessen, a student at the National Center for Supercomputing Applications at the University of Illinois. He and his colleague Eric Bina wrote a new web browser called Mosaic. Although it was originally written for the UNIX-based machines used in academia, Mosaic was then rewritten to run on PCs and Macintosh computers used in homes and offices. Easy to install and use, and able to display graphics embedded in web pages, rather than in a separate on-screen window, Mosaic made the web accessible to the general public for the first time. For newcomers to the Internet struggling with different programs for different functions, Mosaic's simple and unified approach was a godsend. By October 1994 a new magazine called *Wired* was declaring Mosaic to be "well on its way to becoming the world's standard interface," noting that "in the 18 months since it was released, Mosaic has incited a rush of excitement and commercial energy unprecedented in the history of the Net."

Berners-Lee had designed the web to encourage collaboration by scientists, providing a new place for people and ideas to meet, just as they had previously in coffeehouses, at scientific societies, and in the pages of scientific journals. "It was always designed to be a social medium," he says. "Originally I wanted it to be the medium by which I could share ideas with people, so it was very much supposed to be a collaborative medium." By 1993 it was clear that the web's appeal extended far beyond academia, as companies and computer enthusiasts began to take an interest in it as well. In its early days the web lacked the power and depth of other Internet sources, but it made exploring the online world simple and fun, and it was instrumental in bringing the Internet into mainstream use. Berners-Lee himself first realized that the web was being embraced by a wider audience than he had anticipated on a summer day in 1993, when he began using a large color screen. As he was browsing the web, then still in its infancy, he stumbled upon a web-based exhibit of Renaissance art from the Vatican, based on images posted online by the Library of Congress, wrapped up in a few simple web pages by a Dutch programmer. As a colorful illuminated manuscript unfurled on his screen, Berners-Lee recalls, it took his breath away. Not only was it beautiful; it also demonstrated the web's power to enable international collaboration and information sharing.

As the web's ability to put a friendly face on the Internet became apparent, it reawakened a continuing debate within CERN about whether the laboratory should try to make money by licensing technology to industry. Berners-Lee argued strongly against trying to profit from the web, reasoning that unless it remained an open standard, it would end up being just one of several incompatible forms of Internet media, backed by rival companies and organizations. The result would be a balkanized Internet that would hinder the smooth sharing of information. Making the web open and royalty-free, following the example of the Internet protocols that underpinned it, would make it more attractive than any proprietary alternative. Berners-Lee eventually prevailed and on April 30, 1993, his bosses at CERN issued a formal declaration that the web's underlying standards would always be royalty-free. "Without that, it never would have happened," he says.

Instead, the commercial exploitation of the web was left to others, and at the forefront of the emerging Internet boom was Netscape Communications, the company cofounded by Marc Andreessen. Unable to get permission from the University of Illinois to use the Mosaic name, his company launched a new version of the browser under the name Netscape Navigator. It quickly became the world's most widely used web browser as the online population mushroomed, growing from fewer than 5 million mostly academic users in 1991 to 40 million people in 1995, 70 million in 1998, and around 250 million by 2000. Thousands of companies sprang up hoping to exploit the popularity of the new medium by selling goods and services, from books to pet supplies, over the web. Many of these firms floated on the stock market, attracting huge valuations despite the fact that they often had little revenue and made no profit. As the Internet population continued to soar, these companies insisted, profits would follow. The mania ended abruptly with a spectacular crash in early 2000, when the overoptimism of forecasts about people's willingness to spend money online became apparent. Though many of the Internet companies that arose during the so-called "dotcom boom" were swept away, the number of people using the Internet continued to grow rapidly, reaching 500 million in 2001 and 700 million in 2003. And while the appetite for online shopping had been overestimated, there was no doubting the Internet's popularity in another area: as a medium that let anyone publish whatever they wanted to a global audience.

Companies, universities, and governments had started setting up web pages and websites during the 1990s. And so did millions of individuals, attracted by the possibilities of a new publishing environment that was open to all. The option to create a personal website was a standard feature of Internet-service packages from the mid-1990s, but to begin with this meant building web pages from scratch, a fiddly business that deterred all but the most dedicated enthusiasts. Berners-Lee's original browser had been capable of both creating and displaying web pages, without having to grapple with the underlying HTML code, but subsequent browsers concentrated on displaying pages rather than creating them. Things gradually improved, however, as AOL, Tripod, GeoCities, and other firms provided new tools for web-based publishing that were easier to use. The result was tens of millions of personal home pages, with photo galleries, flashing text and graphics, personal journals, and lists of noteworthy links to other sites. Web publishing was then further simplified in the late 1990s with the emergence of new tools designed around an increasingly popular format: the blog.

The word *weblog* was coined in 1997 by Jorn Barger, a programmer and prolific contributor to Usenet, to describe his habit of "logging the web" by publishing a daily log of interesting items on his website, RobotWisdom.com. The term came into wider use when it was abbreviated to "blog" by another programmer, Peter Merholz, in 1999. The word *blog* came to mean the combination of four ingredients, each of which had existed online for several years, so that many sites can be considered, in retrospect, as blog-like or as early blogs. First, a blog consists of a series of personal entries, each of which is dated, as in a journal or diary, and typically contains links to other web pages. Second, these entries are displayed in "reverse-chronological" order, so that visitors to the blog will always see the most recent entry at the top, with older entries beneath it. Third, readers are invited to leave comments beneath each entry, responding both to the author and to each other. Commenting was not a new idea in itself. Web-based discussion sites, which let people leave comments and hold Usenet-style threaded conversations on a single web page, dated back to 1994. Similarly, many people had published daily journals presented in reverse-chronological order before the term *blog* was coined. But the crucial fourth ingredient that made blogging take off was the emergence of easy-to-use, automated publishing tools that provided reverse-chronological entries and commenting as standard features. An important early example was blogger.com, a service launched in 1999 by Evan Williams and Meg Hourihan that made it easy for anyone to set up and run a blog. Williams also popularized the use of the word *blog* as a verb as well as a noun, and of the word *blogger* to refer to someone who blogs.

Blogger.com, Xanga, LiveJournal, and other such sites meant that technical prowess with web-publishing tools became almost irrelevant; what mattered was having something interesting to say and the discipline to maintain a steady flow of blog posts that would attract an audience and elicit responses from readers and other bloggers. Bloggers began providing lists of links to other blogs they recommended, and "trackbacks" allowed bloggers to see when others had linked to their posts. Many conventions from the age of pamphleteering were deliberately or unwittingly revived, such as the use of carefully chosen pseudonyms and the practice of quoting another writer's argument, one paragraph at a time, and then responding to each paragraph in turn. As the number of blogs increased, and bloggers responded to each other's posts with posts of their own, the result was a network of interlinked blogs, surrounded by a sea of comments from readers. Ideas traveled through this network as bloggers and commenters linked to related posts elsewhere. This collective online discussion became known as the "blogosphere."

It is a sign of a medium's immaturity when one of the main topics of discussion is the medium itself. This was true of radio in its early days, and of Usenet, and of the early blogosphere. But in 2000 and 2001 the discussion broadened as blogs covering American politics began to appear, such as AndrewSullivan.com, Talking Points Memo, Instapundit, and Little Green Footballs. In December 2002 political bloggers were instrumental in causing Trent Lott, the U.S. Senate Republican leader, to resign after he made remarks at a party that were taken to indicate his support for the policy of racial segregation. Lott's controversial remarks received very little coverage in the conventional media, but they were seized upon by bloggers who kept the story alive and dug up other examples of Lott saying similar things. This prompted the news media to reexamine the story and give it more coverage. Lott issued an apology in a vain effort to draw a line under the affair, but he ended up resigning a few days later.

As a result of episodes like this, bloggers in America came to see themselves as an alternative, and a corrective, to the mainstream media (commonly referred to as the MSM). And they were indeed its mirror image: decentralized, opinionated, and often fiercely politically partisan. The established news media, for its part, played up another distinction between blogs and mainstream news sources: that blogs were written by amateurs, rather than professional journalists. Bloggers saw this as an advantage, because it meant their medium was open to anyone. Many journalists, however, considered bloggers to be untrustworthy interlopers on their territory, or parasites feeding on the output of traditional media organizations. Writing in November 2004 in response to bloggers' coverage of that year's presidential election, Eric Engberg, a former CBS News correspondent, declared that "given their lack of expertise, standards and, yes, humility, the chances of the bloggers replacing mainstream journalism are about as good as the parasite replacing the dog it fastens on."

The high point of the antagonism between American bloggers and the mainstream media came in late 2004, when *60 Minutes*, an evening news show on CBS, alleged on the basis of leaked memos that President George W. Bush had used family connections to win favorable treatment in the Air National Guard in the 1970s. Little Green Footballs and other blogs immediately questioned the authenticity of the memos, analyzing them in great detail and convincingly demonstrating that the documents, supposedly written in 1972–73, had in fact been created using modern word-processing software. During the resulting furor a former CBS News executive derided blogging as "a guy sitting in his living room in his pajamas, writing what he thinks." But the bloggers were right, and the professional journalists were wrong. Unable to prove that the documents were authentic, CBS retracted the story and Dan Rather, one of the most respected names in American news, resigned as the show's anchor in early 2005.

In the years that followed, the hostility between bloggers and the mainstream media gradually abated. Some prominent bloggers were hired by newspapers or magazines: Andrew Sullivan moved his blog to the website of *Time* magazine in 2006, for example. Many mainstream news organizations launched blogs of their own, written by their journalists (sometimes, in their pajamas), and adopted the blog convention of allowing readers to comment on articles. As they turned into commercial ventures and began to pursue advertising revenue, meanwhile, some blogs became more magazine-like, with web pages showing the most important or popular stories at the top, rather than simply the most recent. Bloggers came to be recognized as valuable and trusted sources in many specialist fields, providing in-depth expertise in areas such as law or science, for example, as well as first-person reporting from countries riven by war or revolution. More broadly, blogging vastly broadened the range of opinions available to online readers. And by holding professional journalists to account, it forced them to raise their game. That is not to say that everything that bloggers write is accurate or worth reading—but the same is true of journalists, as bloggers have shown. As Harold Evans, former editor of the *Times* of London, told the *Wall Street Journal* in 2007, "some blogs have become the best check on monopoly mainstream journalism, and they provide a surprisingly frequent source of initiative reporting."

## WRITING ON THE WALL

It took only a few years for blogging's conventions and publishing tools, which were once thought to exist in a parallel universe to the mainstream media, to become widely accepted. But although blogging made Internet publishing easier than ever, it was still a minority sport. By the end of 2006 there were more than one billion people using the Internet, compared with around forty million blogs. In theory the Internet's openness meant anyone could publish information, but in practice only a small proportion of users chose to do so by blogging. Setting up a blog, posting regularly, and building up a loyal readership still seemed like too much trouble to most people. Instead a new set of publishing tools, built around the sharing of information with family, friends, and other contacts, was to prove far more popular.

Like blogs, social-networking sites began as outgrowths from the personal home pages created in the early days of the web. Some of these pages converged on the reverse-chronological journal format that had given rise to blogs, but another popular type of page was the online résumé or profile page, with biographical details, photos, and a list of interests and favorite links. In the late 1990s several websites let users create profile pages of this nature. SixDegrees.com, which launched in 1997, was the first site to take the next step by allowing users to create lists of their friends. Visitors to any user's page would see a list of their friends, any of which could be clicked to call up their profile in turn, along with their own list of friends. Users of the site could then find out how they were connected (if at all) to another user on the site, and could send messages to their friends, or to anyone known to their friends or to their friends' friends. Although SixDegrees.com signed up around a million users, it was never a commercial success and closed in 2000. But by this time other sites had adopted the idea of allowing users to create and display lists of friends and then traverse the resulting social networks.

The original idea behind Friendster, a website launched in 2002, was to act as a dating website to put users in touch with people with whom they had a friend in common—in other words, friends of their existing friends. The site's users could create an online profile, send messages to people to whom they were connected, and post "status updates" indicating their mood, or what they were doing. Friendster had signed up three hundred thousand users by May 2003, when it started to attract attention from the news media, prompting an influx of millions more users. Friendster's servers struggled to cope with the greater load, irritating users who had become reliant on the site's messaging tools as an alternative to e-mail. One of the things that slowed it down was the constant recalculation of the number of "friends of friends of friends" with whom each user was allowed to communicate. As the number of registered users grew, this calculation took longer to complete, and pages took longer to load. Meanwhile, the company was concentrating on developing new features, rather than on making the site run more smoothly. In 2004 disgruntled users began to decamp en masse to MySpace, a rival social-networking site set up in late 2003 in response to Friendster's early success.

MySpace grew explosively during 2004, quickly moving beyond its initial audience of Friendster refugees and signing up millions of teenagers. Part of its appeal stemmed from the fact that its profile pages were highly customizable and could be made public, whereas profile pages on Friendster were visible only to other registered users of the site. By 2005 MySpace had twenty-five million users, and that year it was sold to News Corp, the media conglomerate owned by Rupert Murdoch, for $580 million. This was regarded as a clever move by Mr. Murdoch, whose media empire, one of the world's largest, had been slow to embrace the Internet. MySpace was, by this time, the fifth most popular site in the United States. In 2006 it reached a hundred million users; the following year it was being valued at twelve billion dollars. But MySpace started running into problems. Its customizable profile pages were vulnerable to attack, hijack, and vandalism and could be configured so that they tried to install malicious software when viewed. The site was beset by spammers, fake profiles, and scantily clad aspiring celebrities, raising concerns about whether it was safe for children. Its new owner, News Corp, treated it as a media outlet rather than a technology platform and seemed more interested in maximizing advertising revenue than in fixing or improving the site's underlying technology. This proved to be a miscalculation. Just as MySpace had overtaken Friendster, MySpace found itself being surpassed in turn by Facebook, yet another new social-networking site.

Mark Zuckerberg, a student at Harvard University, had launched Facebook in 2004, initially for the sole use of Harvard undergraduates, half of whom signed up in the first month. After this initial success he then gradually opened it up to wider use. First, students at other universities were admitted, then high school students and corporate users. Admission depended on having an e-mail address from an approved institution. This exclusivity contributed to Facebook's early appeal, giving it the feel of an elite, private club, just at the time when MySpace was gaining a reputation as a rather lurid and crassly commercial corner of the Internet. Facebook's clean, minimal design was in stark contrast with the garish personalization that was the norm on MySpace. Finally, in September 2006, Facebook was opened to anyone aged thirteen or over. Its ascendancy was apparent long before Facebook overtook MySpace in registered users in early 2009.

Like previous social-networking sites, Facebook let users browse lists of their friends' friends, post messages on each other's profile pages (each of which, in Facebook parlance, was called a Wall), and indicate what they were doing by posting status updates. But Facebook did a better job of moving with the times, adding new features to make pages more responsive, allowing other companies to develop games and other software to run on Facebook pages, and making Facebook features, such as the "Like" button, which indicates approval, available on websites other than its own. Perhaps most importantly, in 2006 Facebook introduced a "news feed" giving users a reverse-chronological list of all of their friends' recent activity on the site: messages posted, status updates, changes to profile information, new links or photos posted, and announcements when new friendships were established or relationships ended. This provided a much quicker way to see what was going on than visiting friends' profile pages in turn, and it made the site amazingly compelling by providing a personalized and constantly updated stream of links, photos, and gossip.

Facebook has since gone on to become the world's largest social-networking site, with more than one 1 billion users. When the company went public in May 2012, it was valued at more than one hundred billion dollars, although its market capitalization quickly fell to around half that sum. Its popularity means it is also the world's largest photo-sharing site, with hundreds of millions of photos being uploaded each day. Such is Facebook's dominance that it accounts for one minute in seven spent online. A study in December 2012 found that it was the most popular social network in 127 of 137 countries analyzed. But the brief history of social-networking sites suggests that it cannot afford to be complacent. In particular, Facebook's repeated changes to its terms of use, many of which have the effect of making more of the information posted on the site publically accessible, are a constant source of controversy and complaint. Unlike with Friendster or MySpace, however, the grumbling about Facebook has not led to a mass exodus of users to a rival site.

The stream-based display of updates, popularized by Facebook's news feed, spawned another approach to the social exchange of information, known as "microblogging," in which brief messages and status updates from friends and contacts are displayed in a reverse-chronological feed. This model was pioneered by Twitter, a start-up cofounded in 2006 by Evan Williams, one of the founders of Blogger. It began when Jack Dorsey, another of Twitter's cofounders, realized that the status updates commonly posted on social networks and instant-messaging systems would be much more useful and informative if people could update them when they were away from their computers. The first incarnation of Twitter was based on mobile-phone text messaging, rather than the Internet. Users could change their current status ("Tom is going to the library") and receive updates from their friends via text messages. For technical reasons text messages are limited to one hundred and sixty characters in length, of which twenty characters were reserved for the user name and routing information, leaving one hundred and forty characters for each message, or "tweet." This limitation, forcing people to be concise, was kept in place when Twitter was launched as an Internet-based service in July 2006. It took off in 2007 after being embraced by the influential and tech-savvy attendees at South by Southwest Interactive, a technology conference. As the volume of tweets grew in the following months, Twitter users developed their own conventions—such as "retweets" (reposting another user's tweet to one's own followers, with attribution) and "hashtags" (labels associating a tweet with a given topic)—that were then officially incorporated into Twitter. As on other social-networking services, noteworthy messages, links, photos, and other items posted by users could often reach a vast audience as users reshared them to their own lists of followers, who reshared them in turn.

By 2013 Twitter had more than two hundred million active users who were collectively sending four hundred million tweets every day. Twitter users include many heads of state, celebrities, and even Pope Benedict XVI, who joined Twitter in December 2012. Although Twitter can be confusing for newcomers, with its conventions and jargon, it provides an appealing mixture of public and private conversations on all kinds of topics, rather like a seventeenth-century coffeehouse. In a speech at the Gerald R. Ford School of Public Policy in November 2012, Dick Costolo, Twitter's chief executive, likened it to a global town square, a reinvention of the Roman Forum or Greek Agora in which people can exchange views directly, rather than getting information from a small number of centralized, filtered sources such as newspapers, radio, and television.

> So interesting things start to happen as technology improves and broadcast media becomes more prevalent. One, as we start to create the ability to create these print broadsheets that can be distributed to more and more and more people, it also increases the cost and capital requirements with being a publisher, right? . . . It was something that fewer and fewer people or companies or organizations had access to. So not only is it now not truly multi-perspective, it's the perspective that's coming from fewer and fewer and fewer sources, ever more filtered. And with that filtering and the fewer sources we start to get really this outside-in view of the news. What I mean by that is in the Agora we had an inside-out view of the news. It was the participants themselves who would come there and talk about what was going on with them or what they had just witnessed or what they just saw . . . But along comes Twitter, and Twitter reinvents the Agora.

With its brief messages, easily sent and received on mobile devices, Twitter offers greater immediacy and a more conversational quality than blogs, e-mail, or exchanges of Facebook messages. It provides a forum for live discussion and running commentary on everything from television shows and sporting events to demonstrations and civil wars. By choosing who to follow, Twitter users can assemble their own personal feeds of information from friends, celebrities, and institutions such as government agencies, companies, and newspapers. The microblogging approach pioneered by Twitter has proved particularly popular in China, where Twitter itself is banned, but where similar microblogging services, called *weibo*, have been set up by local Internet firms. Tencent Weibo, Sina Weibo, and Netease Weibo all have hundreds of millions of users, with more than half of China's six hundred million Internet users using at least one of them. Microblogging works particularly well in China because smartphones (rather than PCs) are the most popular Internet-access devices. And Chinese allows many words to be expressed using very few characters, making it possible to fit an entire paragraph into a single weibo post.

Facebook is also banned in China, where a similar site called QZone has more than half a billion users. Even in some countries where Facebook and Twitter are readily accessible, other sites are more popular:

QZone in South Korea and VKontakte in Russia, for example. The rise of MySpace, Facebook, and other social networks led many media-sharing sites, such as YouTube and Flickr, to add social features, and prompted the launch of countless variations on social networking and social media. Tumblr, a service launched in 2007 that sits midway between Twitter and a full blogging platform, has around two hundred million users. Google, the Internet giant that dominates online search and advertising, launched its own social network, called Google+, in 2010. Instagram, a photo-sharing take on Twitter, was acquired by Facebook in 2012 for more than seven hundred million dollars. Pinterest lets users collect and share images in digital scrapbooks, or pinboards. Path is a mobile-only social network that limits users to the "Dunbar number" of one hundred and fifty friends. Medium, the latest venture from Evan Williams, is akin to both a magazine and a shared blogging platform. Branch is a discussion platform that allows more structured exchanges of longer posts than Twitter does. Every week, it seems, a new variation on social networking, sharing, and publishing appears. Given how widely used social sites have become, it is hardly surprising that, like earlier forms of social media, they have started to have social and political impact, particularly in countries where publishing has traditionally been tightly restricted.

## SOCIAL MEDIA IN THE ARAB SPRING AND BEYOND

On December 17, 2010, a twenty-six-year-old Tunisian fruit seller, Mohamed Bouazizi, had his produce and weighing scales confiscated by police in his home town of Sidi Bouzid. Exasperated by repeated harassment and insults from officials, he went to the regional governor's office to complain. When the governor refused to see him, Bouazizi doused himself with fuel, cried out, "How do you expect me to make a living?" and set himself on fire. His protest prompted immediate demonstrations by other street vendors in Sidi Bouzid, and a large crowd gathered outside the governor's office. A peaceful protest the next day was broken up by police using tear gas. State-run media made no mention of Bouazizi's action, but pictures of him in his hospital bed, and mobile-phone footage of the protests, began spreading on Facebook, YouTube, blogs, and other social sites.

One such video was seen on Facebook by journalists at Al Jazeera, a satellite-news broadcaster based in Qatar and the Arab world's most influential media outlet. Al Jazeera showed the video on air, giving it not just national but international prominence. Other satellite broadcasters followed suit, showing images and video of protests, taken from social media sites, as the unrest rippled across Tunisia. This made them available even to the two thirds of Tunisia's population without Internet access. Previous protests, such as that in the town of Ben Guerdane a few months earlier, had been quickly and brutally stifled by the police, preventing news of any unrest from spreading. But this time was different. By the time Bouazizi died of his burns on January 4, 2011, antigovernment protests had broken out across the country.

The Tunisian authorities tried to block Facebook, Twitter, and video-sharing sites such as YouTube and Dailymotion. But the protests had become too widespread and numerous to be suppressed. The potent combination of social media and satellite television revealed to young Tunisians, frustrated by corruption, high food prices, unemployment, and a lack of political freedom, that their views were widely shared, emboldening them to take collective action. Bouazizi's cousin Ali, who, along with another family member, had recorded the first protest on his mobile phone, later explained, "We tried to record everything that happened, and to broadcast it through the Internet, because Tunisian media always silenced us. We were the first to send these pictures, which had an important echo around the world." President Zine El Abidine Ben Ali declared a state of emergency, dissolved the government, and promised new elections, but the protests continued. On January 14 Ben Ali fled the country, bringing to an end his twenty-three-year rule.

The success of the Tunisian uprising inspired similar protests across the Arab world. In Egypt, as in Tunisia, there was a thriving underground network of young activists well versed in the use of blogs and other forms of social media and connected to a foreign diaspora of sympathizers. In 2010 they had helped spread images of Khaled Said, a twenty-eight-year-old Egyptian who had been beaten to death by two policemen, apparently for posting online a video of corrupt police officers engaged in a drug deal. The initial police report claimed that Said had suffocated after taking drugs, but a member of his family obtained a mobile-phone picture of his badly beaten body and shared it online, causing an outcry. Protests were held in Cairo and Alexandria, and Said's death prompted Wael Ghonim, a twenty-nine-year-old Egyptian who worked for Google in Dubai, to create a Facebook page called "We Are All Khaled Said." Created under the pseudonym "ElShaheed" ("The martyr"), the page had soon attracted 350,000 followers, in Egypt and beyond. "I wanted to tell people that the same could happen to me, or to anyone else," Ghonim told an Egyptian newspaper via an anonymous Internet call. The page became the coordination point for several silent protests by Said's supporters in late 2010, during which they dressed in black and took to the streets. They stood silently together to avoid falling foul of Egypt's emergency law, which banned the chanting of slogans. The page also advocated writing "No to Torture, No to the Emergency Law" on Egyptian banknotes, so that this slogan would circulate as money passed from hand to hand.

After the fall of Tunisia's president on January 14, Ghonim posted a message on the page inviting Egyptians to take to the streets on January 25, a forthcoming national holiday. The page became a rallying point for activist groups, bringing together a number of calls for a day of antigovernment protests. Ghonim returned to Egypt so that he could take part in the demonstration himself. The nationwide protests that followed, coordinated in part via mobile phones and social media, prompted the government to shut down the country's communications networks entirely on January 26. This was a mistake, Ghonim says, because it showed that the government was scared. "Fear was the main reason that the regime was still going. But when you cut the Internet you are telling everyone you are scared. So people took to the streets because they thought: we are stronger."

Days of protest followed, during which Egypt's president, Hosni Mubarak, vainly offered concessions, but refused to resign. Ghonim disappeared but then emerged as one of the figureheads of the protest movement when he gave an emotional interview on Egyptian television after being released from twelve days of police captivity, helping to revitalize the protest movement just as it seemed to be losing momentum.

Eventually, after two weeks of continuous protest against his rule, Mubarak resigned on February 11, 2011. Speaking in 2011, Naguib Sawiris, an Egyptian telecommunications billionaire who backed the uprising, applauded "what this technology has done to my part of the world. Ninety percent of the success of this revolution is attributed to it." As in Tunisia, Egypt's revolution was hailed in parts of the Western media as a "Facebook revolution." That is going a bit far, but Facebook was among the forms of social media that had once again played a part in spurring a revolution.

Both optimists and pessimists about social media's potential to bring about political change are now watching China closely. It is the world's most populous country, with the largest number of Internet users, hundreds of millions of whom are using weibo. But China also has the world's most advanced Internet-censorship system. Operators of websites and Internet services are required to delete or suppress posts containing keywords relating to sensitive subjects, such as the names of senior politicians and human-rights activists. Other terms can be banned when necessary. In October 2012, for example, when the *New York Times* published an investigation into the wealth accumulated by Wen Jiabao's family during his term as prime minister, the authorities banned all weibo posts containing "New York Times," "NYT," and the names of several of Mr. Wen's relatives. A study of weibo censorship by Dan Wallach and his colleagues at Rice University in Houston, Texas, published in March 2013, concluded that there were more than four thousand censors working around the clock in eight-hour shifts to monitor the seventy thousand weibo messages posted every minute. The total number of people involved in the internet-censorship system, known as "Golden Shield," is estimated to be around one hundred thousand.

Arguably weibo and other online forums provide a way for people to let off steam, while also giving the government a handy means of keeping track of public opinion, as was the case for manuscript news networks of seventeenth-century England and for the poems that circulated in prerevolutionary France. The *People's Daily Online*, an affiliate of the Communist party's main newspaper, analyzes social media to compile a weekly report called "Online Public Sentiment," which is available only to senior officials. In 2008 Hu Jintao, then president of China, told the *People's Daily* that "the web is an important channel for us to understand the concerns of the public and assemble the wisdom of the public." And despite the constraints on the use of weibo and other Internet forums, the toleration of political discussion within certain limits has undoubtedly given people a new means by which to express their concerns and has made it harder for the government to conceal or ignore incompetence and corruption. A striking example came in the aftermath of a high-speed train crash near the city of Wenzhou in July 2011, in which forty people died and 191 were injured. At first the government blamed the accident on signaling problems caused by a lightning storm, and literally tried to cover it up. Carriages from the crashed train were broken up and buried within hours of the crash in an effort to protect the reputation of China's railway industry, a government favorite. But details of the incident, including images of carriages being buried, were soon circulating on weibo.

The resulting outcry and the widespread ridicule by Internet users of the railway ministry's feeble explanations—it claimed, for example, that the carriages had to be buried to facilitate the rescue work, and to prevent industrial espionage by foreigners—prompted the government to change its tune. Five days after the crash the prime minister, Wen Jiabao, belatedly visited the scene, criticized the railway ministry, and ordered a full inquiry. Several officials in the railway ministry lost their jobs when the cause of the accident was found to be a combination of management failures and faulty signaling systems. But Internet posts relating to the crash were closely policed to prevent the outburst of anger it had prompted from turning into to broader criticism of the government, and to prevent the organization of offline protests. The advent of weibo has given Chinese users a newfound freedom to express themselves, but within limits that are still carefully circumscribed by the government.

The Wenzhou train crash also illustrates how online criticism of the government in China differs from that in Arab countries. In Tunisia and Egypt, young people who lacked jobs and economic opportunity, and felt they had nothing to lose, led the calls for the removal of their despotic leaders. In China, by contrast, middle-class Internet users are worried that the wealth they have accumulated during China's rapid economic development is in danger of appropriation by corrupt officials and a capricious and inconsistent application of the law. Their political engagement is a consequence of wealth, not poverty. As a result, the smartphone-toting Chinese middle class and the ruling elite have, to some extent, a common interest in a gradual process of steady reform, rather than a cataclysmic political upheaval.

From the earliest mailing lists to the latest microblogging platforms, the history of the Internet has involved the progressive lowering of barriers to being able to publish. Twitter and its ilk reduce the amount of effort needed to publish something almost to zero: a few taps on a smartphone's screen can fire off words or pictures with the potential to be seen by an audience of millions. A study by Denis Pelli of New York University and Charles Bigelow of the Rochester Institute of Technology, published in 2009, considered how the number of published authors per year, as a proportion of the total population, has increased since the advent of the printing press. Between 1500 and 2000, they found, the number of published authors per year (defined as those able to reach an audience of one hundred or more readers) increased from around one hundred to around one million, or less than 0.01 percent of the population. The advent of blogs, social-networking sites, microblogs, and media-sharing services since the turn of the twenty-first century has increased that figure to more than 1.5 billion, or about 20 percent of the world population. "Authors, once a select minority, will soon be a majority," note Pelli and Bigelow. Today's social media can be seen as the culmination of a long historical trend that has gradually extended the ability to share one's thoughts with others to a larger and larger proportion of the population. It seems likely that it will not be many years before the whole world has access to the social publishing medium of the Internet.

After a one-hundred-and-fifty-year hiatus during which the person-to-person aspect of media was overshadowed by centralized mass media operating on a broadcast model, the pendulum has swung back. Social forms of media based on sharing, copying, and personal recommendation, which prevailed for centuries, have been dramatically reborn, supercharged by the Internet. By making it quick and easy for anyone to share information with others, modern social media gives ordinary people a collective agenda-setting power that was previously restricted to large publishers and broadcasters, and that is capable of striking fear into those in authority. Working out the implications and long-term consequences of this new media environment is the giant collective experiment in which humanity is now engaged. We are still only at the very beginning of the process. But history can provide some helpful clues.

# Epilogue: History Retweets Itself

*The longer you can look back, the further you can look forward.*
—Winston Churchill, 1944

## learning from "really old" media

IN THE YEARS since the Internet became widespread, it has become commonplace to draw a distinction between "new" media based on digital technologies and the "old" media that came before it. But old media, it is now apparent, was something of a historical anomaly. It began in 1833 with the launch of the New York *Sun*, with its innovative mass-media model based on amassing a large audience and then selling their attention to advertisers. Look back before 1833 to the centuries before the era of old media began, however—to what could be termed the era of "really old" media—and the media environment, based on distribution of information from person to person along social networks, has many similarities with today's world. In many respects twenty-first-century Internet media has more in common with seventeenth-century pamphlets or eighteenth-century coffeehouses than with nineteenth-century newspapers or twentieth-century radio and television. New media is very different from old media, in short, but has much in common with "really old" media. The intervening old-media era was a temporary state of affairs, rather than the natural order of things. After this brief interlude—what might be called a mass-media parenthesis—media is now returning to something similar to its preindustrial form.

Admittedly, the analogy between ancient (analog) and modern (digital) forms of social media is not perfect, and there are several important differences. As author and Internet scholar Clay Shirky of New York University has pointed out, Internet-based publishing is instant, global, permanent, and searchable in a way that earlier forms of social media such as papyrus rolls, poems, and pamphlets were not. But the historical forms of social media have enough in common with the modern sort—in their underlying social mechanisms, the reactions they provoked, and the impact they had on society—that they can help us reassess social media today, and the contemporary debates it has engendered.

The most obvious area where history can provide a helpful perspective is in the debate over the political impact of social media and its role in triggering protests and revolutions. This has been a hotly contested subject for some time—the Arab Spring has only intensified the argument. On the one hand are those, like Shirky, who point to the use of social media by activists and revolutionary movements, notably in Tunisia and Egypt, but also elsewhere; on the other are those, like the writers Malcolm Gladwell and Evgeny Morozov, who have expressed skepticism that online support for a cause will necessarily translate into real-world action. Indeed, skeptics argue, supporting a cause online may even make people less likely to take action, because they may then feel they have done their bit (what Morozov likes to call "slacktivism"). Why go on a march in favor of Tibetan self-determination, say, when you can just click a "Like" button on Facebook instead?

Historically, it is clear that social media, in the form of pamphlets, letters, and local newspapers, played a role in the Reformation, and in the American and French revolutions. But it is also clear, from a distance, that its main function was to reveal and synchronize public opinion and expose the extent of the opposition to the incumbent regime. In each case, simmering resentments meant that revolution would have happened sooner or later anyway; the use of social media merely helped the process along. It is wrong to attribute revolutions entirely to social media, in other words, but it is also wrong to discount its role altogether.

A helpful perspective, common to the Arab Spring and the

Reformation, is that getting a revolution going is like starting a fire. Jared Cohen, a former official in America's State Department who now works for Google, has likened social media's role in the Arab Spring to that of an "accelerant" that causes fire to spread more quickly. A similar view was expressed in an illuminated manuscript from 1572, showing how the tinderbox of European religious discontent was finally ignited. John Wycliffe is depicted holding a match, Jan Hus a candle, and Martin Luther a flaming torch. New forms of social media did not start the fire in either the sixteenth or twenty-first centuries. But in both cases it helped turn an initial spark into a conflagration.

A related question is whether greater access to the Internet, and to the open publishing environment provided by social media in particular, inherently promotes freedom and democracy. This view was most clearly stated by Wael Ghonim after the 2011 uprising in Egypt. "I always said that, if you want to liberate a society . . . if you want to have a free society, just give them the Internet," he told CNN. Asked which country would follow Tunisia and Egypt in overthrowing its government, Ghonim replied, "Ask Facebook."

This is reminiscent of Condorcet's utopian suggestion, in the 1790s, that "the press . . . has freed the instruction of the people from every political and religious chain." Greater freedom of expression certainly makes life more difficult for repressive regimes. But as events in Condorcet's time showed, unfettered publishing also makes it easier for governments to monitor public opinion and keep tabs on dissidents; and as the aftermath of the French Revolution demonstrated, in a lawless environment a free press can be exploited by populists, leading to mob rule.

In the modern era, similar concerns about the drawbacks of social media have been expressed by Morozov, who notes that oppressive regimes can exploit it in several ways. In particular, it helps governments spread propaganda and makes possible new forms of surveillance, by making it easier to identify connections between activists. In Morozov's birthplace Belarus, for example, where dissidents used the LiveJournal blogging platform to coordinate their actions, "social media created a digital panopticon that thwarted the revolution: its networks, transmitting public fear, were infiltrated and hopelessly outgunned by the power of the state." He concludes that "the emergence of new digital spaces for dissent also leads to new ways of tracking it . . . getting access to an activist's inbox puts all their interlocutors in the frame, too."

That said, social media, whether in the form of the printing press or the Internet, can be a force for freedom and openness, simply because oppressive regimes often rely on manipulating their citizens' view of the world, and a more open media environment makes that harder to accomplish. But the other side of the scales is not empty; this benefit must be weighed against the fact that social media can make repression easier, too. As Morozov notes, the Internet "penetrates and reshapes all walks of political life, not just the ones conducive to democratization." Anyone who hopes that the Internet will spread Western-style liberal democracy must bear in mind that the same digital tools have also been embraced by campaigners with very different aims, such as Hezbollah in Lebanon and ultra-right-wing nationalist groups in Russia. The test case in this argument is China, which now has more Internet users than any other country—more than in North America and Europe combined. Weibo and other online forums have given Chinese Internet users unprecedented freedom to express their views. Yet the swift and ruthless censoring of blog posts and weibo messages criticizing senior officials or calling for real-world demonstrations shows that widespread Internet adoption need not necessarily threaten the regime. Indeed, the ability to monitor the Internet may make it easier for the government to keep the lid on dissent.

A rather more mundane but widely expressed concern about social media is that the ease with which anyone can now publish his or her views online, whether on Twitter, on blogs, or in comment threads, has led to a coarsening of public discourse. Racism, sexism, bigotry, incivility, and ignorance abound in many online discussion forums. Twitter allows anyone to send threats or abuse directly to other users. No wonder the Internet is often likened to a sewer by politicians, clergymen, and newspaper columnists.

Yet the history of media shows that this is just the modern incarnation of the timeless complaint of the intellectual elite, every time technology makes publishing easier, that the wrong sort of people will use it to publish the wrong sorts of things. In the early sixteenth century, Erasmus complained that printers "fill the world with pamphlets and books that are foolish, ignorant, malignant, libelous, mad, impious and subversive; and such is the flood that even things that might have done some good lose all their goodness." Worse, these "swarms of new books" were "hurtful to scholarship" because they lured readers away from the classics, which is what Erasmus felt people ought to have been reading.

Printers had, however, quickly realized that there was a far larger audience, and more money to be made, printing pamphlets and contemporary works rather than new editions of classical works. Similarly, in England, the Worshipful Company of Stationers bemoaned the explosion of unlicensed pamphlets that appeared after the collapse of press controls in 1641, complaining that "every ignorant person that takes advantage of a loose presse may publish the fancies of every idle brain as so manyfestly appeareth by the swarmes of scandalous and irksome pamphletts that are cryed about the streetes." The Company was hoping to be granted a renewed monopoly on printing, which had previously allowed it to control what was printed, and therefore what people read. Its grumbling is not dissimilar to that of professional journalists bemoaning the rise of pajama-clad bloggers, invading their turf and challenging the status quo.

Those in authority always squawk, it seems, when access to publishing is broadened. Greater freedom of expression, as John Milton noted in *Areopagitica*, means that bad ideas will proliferate as well as good ones, but it also means that bad ideas are more likely to be challenged. Better to provide an outlet for bigotry and prejudice, so they can be argued against and addressed, than to pretend that such views, and the people who hold them, do not exist. In a world where almost anyone can publish his or her views, the alternative, which is to restrict freedom of expression, is surely worse. As Milton's contemporary Henry Robinson put it in 1644, "It were better that many false doctrines were published, especially with a good intention and out of weaknesse only, than that one sound truth should be forcibly smothered or wilfully concealed; and by the incongruities and absurdities which accompany erroneous and unsound doctrines, the truth appears still more glorious, and wins others to the love thereof." One man's coarsening of discourse is another man's democratization of publishing. The genie is out of the bottle. Let truth and falsehood grapple!

Whatever you think about the standards of online discussions, there is no doubt that people are spending a lot of time engaging in them. This raises another concern: that social media is a distracting waste of time that diverts people from more worthwhile pursuits, such as work and study. Surveys carried out in 2009 found that more than half of British and American companies had banned workers from using Twitter, Facebook, and other social sites. Many employers also block access to LinkedIn, a social-networking site for business users, because they worry that it allows employees to spend their time networking and advertising themselves to other potential employers. Simply put, companies readily equate social networking with social notworking.

This too is a familiar worry. Coffeehouses, the social-media platforms of their day, inspired similar reactions in the seventeenth century. They were denounced in the 1670s as "a vast loss of time grown out of a pure novelty" and "great enemies to diligence and industry." But the mixing of people and ideas that occurred in coffeehouses, where patrons from many walks of life would gather to discuss the latest pamphlets, led to innovations in science, commerce, and finance. By providing an environment in which unexpected connections could be made, coffeehouses proved to be hotbeds of collaborative innovation.

Similarly, a growing number of companies have concluded that social networking does have a role to play in the workplace, if done in the right way. They have set up "enterprise social networks," which create a private, Facebook-like social network to facilitate communication among employees and, in some cases, with workers at client and supplier companies, too. This sort of approach seems to have several benefits: its similarity to Facebook means little or no training is required; sharing documents and communicating via discussion threads is more efficient than using e-mail; it is easier to discover employees' hidden knowledge and talents; and it makes it easier for far-flung teams to collaborate.

A study by McKinsey and Company, a management consulting firm, found that the use of social networking within companies could increase the productivity of skilled knowledge workers by 20 to 25 percent and that the adoption of the technology in four industries (consumer goods, financial services, professional services, and advanced manufacturing) could create economic benefits worth between $900 billion and $1.3 trillion a year. Such predictions should always be taken with a very large dose of salt, but McKinsey found that 70 percent of companies were already using social technologies to some extent, and more than 90 percent said they were already benefitting as a result. Far from being a waste of time, then, Facebook-like social networks may in fact be the future of business software.

Even if it has value in the office, however, is there a danger that social media is harming our personal lives? Some observers worry that social media is in fact antisocial, because it encourages people to commune with people they barely know online to the detriment of real-life relationships with family and friends. "Does virtual intimacy degrade our experience of the other kind and, indeed, of all encounters, of any kind?" writes Sherry Turkle, an academic at MIT, in her book *Alone Together*. She worries that "relentless connection leads to a new solitude. We turn to new technology to fill the void, but as technology ramps up, our emotional lives ramp down." Similarly, William Powers, author of *Hamlet's BlackBerry*, laments the way that his family would rather chat with their online friends than with each other. "The digital crowd has a way of elbowing its way into everything, to the point where a family can't sit in a room together for half an hour without somebody, or everybody, peeling off," he writes. His proposed solution: an "Unplugged Sunday" when the use of computers and smartphones is banned.

It is clear that the desire to be connected to one's distant friends, using whatever technology is available, is timeless. Cicero particularly valued the way letters connected him to his friends in the months after the death of his beloved daughter Tullia in 45 B.C. And he relished the contact his daily letters with his friend Atticus provided, even when they contained little information. "Write to me . . . every day," he wrote to Atticus. "When you have nothing to say, why, say just that!" Concerns about unhealthy dependence on new media technologies also have a long history: recall Plato's objections to writing in the *Phaedrus*, and Seneca's derision of his fellow Romans as they rushed to the docks to get their mail. By the seventeenth century, satirists were lampooning news junkies and the hunger with which they sought out the latest corantos.

From Roman letter-writers to manuscript poetry-sharing networks to news-sharing clergymen in the American colonies, the exchange of media has long been used to reinforce social connections. The same is true today. Zeynep Tufekci, a media theorist at Princeton University, suggests that the popularity of social media stems from its ability to reconnect people in a world of suburbanization, long working hours, and families scattered around the globe by migration. Social media, she argues, is also a welcome antidote to the lonely, one-way medium of television. People who use social media can stay in contact with people they would otherwise lose touch with and make contact with like-minded individuals they might otherwise have never met. "Social media is enhancing human connectivity as people can converse in ways that were once not possible," Tufekci argues. A study published in 2011 by researchers at the University of Pennsylvania concluded that "it is incorrect to maintain that the Internet benefits distant relationships at the expense of local ties. The Internet affords personal connections at extreme distances but also provides the opportunity for new and supplemental local interaction." Another analysis, conducted in 2009 by researchers at the University of Toronto and involving four thousand Canadians, found that 35 percent felt that technology made them feel closer and more connected to other family members, and only 7 percent said that technology made them feel less connected. Tellingly, 51 percent of respondents said it made no difference, which suggests that many people no longer make a distinction between online and offline worlds, but regard them as an integrated whole.

New technologies are often regarded with suspicion. Turkle worries about the "flight from conversation," citing teenagers who would rather send a text than make a phone call. And on Unplugged Sunday, Powers and his family engage in communal pursuits that include watching television together. It seems odd to venerate the older technologies of the telephone and the television, though, given that they were once condemned for being anti-social in the same way social media is denounced today. ("Does the telephone make men more active or more lazy? Does it break up home life and the old practice of visiting friends?" asked a survey carried out in San Francisco in 1926.) There is always an adjustment period when new technologies appear, as societies work out the appropriate etiquette for their use and technologies are modified in response. During this transitional phase, which takes years or even decades, technologies are often criticized for disrupting existing ways of doing things. But the technology that is demonized today may end up being regarded as wholesome and traditional tomorrow, by which time another apparently dangerous new invention will be causing the same concerns.

What clues can history provide about the future evolution of social media? Even though Facebook, Twitter, and other social platforms provide a way for people to share information by sharing along social connections, they still resemble old-fashioned media companies such as newspapers and broadcasters in two ways: they are centralized (even though the distribution of information is carried out by the users, rather than the platform owners) and they rely on advertising for the majority of their revenue. Centralization grants enormous power to the owners of social platforms, giving them the ability to suspend or delete users' accounts and censor information if they choose to do so—or are compelled to do so by governments. Relying on advertising revenue, meanwhile, means platform owners must keep both advertisers and users happy, even though their interests do not always align. As they try to keep users within the bounds of their particular platforms, to maximize their audience for advertising, the companies that operate social networks have started to impose restrictions on what their customers can do and on how easily information can be moved from one social platform to another. In their early days, it makes sense for new social platforms to be as open as possible, to attract a large number of users. Having done so, however, such platforms often try to fence their users into "walled gardens" as they start trying to make money.

The contrast between big social platforms on the one hand, and e-mail and the web on the other, is striking. Both e-mail and web publishing work in an entirely open, decentralized way. The servers that store and deliver e-mail and the programs used to read and write messages are all expected to work seamlessly with each other, and for the most part they do. The same is true of web servers, which store and deliver pages, and the web browsers used to display pages and navigate between them. Anyone who wants to set up a new e-mail or web server can add it to the Internet's existing ecosystem of such servers. If you are setting up a new blog or website, there are also plenty of companies to choose from who will host it for you, and you can move from one to another if you are unsatisfied with their service. None of this is true for social networking, however, which takes place inside huge, proprietary silos owned by private companies. Moving your photos, your list of friends, or your archive of posts from one service to another is difficult at best, and impossible at worst. It may be that healthy competition among those companies, and a reluctance to alienate their hundreds of millions of users by becoming too closed, will enable the big social platforms to continue in this semi-open state for many years to come.

But another possibility is that today's social platforms represent a transitional stage, like AOL and CompuServe in the 1990s. They were proprietary, centralized services that introduced millions of people to the wonders of the Internet, but they were eventually swept aside by the open web. Similarly, perhaps the core features of social networking and social media—maintaining lists of friends, and exchanging information with them—will move to an open, decentralized model. Such a model is possible for e-mail and web publishing because of the existence of agreed technical standards on how e-mail messages and web pages ought to be encoded and transmitted. Several such standards have already been proposed for decentralized or distributed social networks, though none has yet gained much traction. There will be technical difficulties synchronizing friend lists, maintaining privacy and security, and delivering updates quickly across millions of users, all of which give centralized social networks a clear advantage at the moment. But every time a major social network is involved in a privacy violation, an unpopular change in the terms of service, or a spat over censorship, a few more adventurous users decide to give one of the various decentralized social networks a try. "I think it's important to design new systems that work in a distributed way," says Tim Berners-Lee. "We must make systems in which people can collaborate together, but do it in a way that's decentralized, so it's not based on one central hub."

A decentralized social platform could be based around personal silos of data over which users would have direct control. This approach would also address concerns that the new online public sphere that has been brought into being by social media is largely in the hands of private companies who are beholden to advertisers and shareholders rather than users. But there is another way for Facebook, Twitter, and other platforms to make themselves more accountable to users and less dependent on advertisers: to start charging users for some or all services. Many Internet services operate on a model in which a small percentage of paying customers subsidize a much larger number of nonpaying users. Social platforms could charge for things such as providing detailed analytics to commercial users of their platforms, more customization options for user profiles, or an advertising-free service. App.net, a subscription-funded Twitter-like service launched in

September 2012, prides itself on being an "ad-free social network" that is based on "selling our product, not our users." This ensures, the company says, that its financial incentives are aligned with those of its members. Whether or not its particular model proves to have broad appeal, the future of social media is likely to see new models based on decentralized architectures and paying customers being added to the mix.

But whatever form social media takes in the future, one thing is clear: it is not going away. As this book has argued, social media is not new. It has been around for centuries. Today, blogs are the new pamphlets. Microblogs and online social networks are the new coffeehouses. Media-sharing sites are the new commonplace books. They are all shared, social platforms that enable ideas to travel from one person to another, rippling through networks of people connected by social bonds, rather than having to squeeze through the privileged bottleneck of broadcast media. The rebirth of social media in the Internet age represents a profound shift—and a return, in many respects, to the way things used to be.

# Acknowledgments

In retrospect it seems quite obvious that having written histories of food and drink, I should go on to write another book on something else you share with your friends, namely media. But it did not seem quite that straightforward when I started, and I am very grateful to all those who helped me develop my thesis, provided leads and suggestions, or shared their opinions in interviews. In particular, I would like to thank Craig Newmark, An Xiao Mina, Jay Rosen, Henry Jenkins, Vint Cerf, Tim Berners-Lee, Wael Ghonim, Matt Locke, and Andrew Lintott. John Micklethwait, Emma Duncan, Ann Wroe, Oliver Morton, Rob Gifford, and Gady Epstein, all colleagues at the *Economist*, provided help of various kinds along the way. I am also grateful to George Gibson, Jackie Johnson, and Katinka Matson for their continued support and encouragement throughout the writing process. Finally I would like to thank my wife, Kirstin, my daughter, Ella, and my son, Miles, to whom this book is dedicated.

# NOTES

## CHAPTER 1: THE ANCIENT FOUNDATIONS OF SOCIAL MEDIA

Statistics for social-media use come from comScore and the Pew Research Center's Internet & American Life Project's "State of Social Media: 2011" report. For the relationship between neocortex ratio and group size, see Dunbar, "Neocortex Size As a Constraint on Group Size in Primates" and "Coevolution of Neocortical Size, Group Size and Language in Humans." The link between neocortex size and deception rate is described by Byrne and Corp, "Neocortex Size Predicts Deception Rate in Primates." The use of gossip as a form of grooming among humans and its possible role in the origin of language is discussed by Dunbar, *Grooming, Gossip and the Evolution of Language* and "Gossip in Evolutionary Perspective"; Foster, "Research on Gossip"; McAndrew, "The Science of Gossip"; and Mithen, *The Prehistory of the Mind.* The account of the origin of writing follows Schmandt-Besserat, *How Writing Came About.* Early literacy and Greek attitudes toward writing are discussed in Harris, *Ancient Literacy*; Ong, *Orality and Literacy*; Kaestle, "The History of Literacy and the History of Readers"; Gibson, "Epilogue to Plato"; and Lewis, *News and Society in the Greek Polis.*

## CHAPTER 2: THE ROMAN MEDIA

The account of Roman letter-writing practice draws upon Edwards, "Epistolography"; Stowers, "Letter Writing in Greco-Roman Antiquity"; Druckenmiller, *Cicero's Letters and Roman Epistolary Etiquette*; Pauli, "Letters of Caesar and Cicero to Each Other"; Mierow, "Julius Caesar as a Man of Letters"; and Cicero, *Letters to Friends* and *Letters to Atticus.* The acta diurna is discussed by Boissier, "Tacitus"; Giffard, "Ancient Rome's Daily Gazette"; and Baldwin, "The Acta Diurna." The best account of the Roman book-publishing scene is to be found in Winsbury, *The Roman Book*. The description and analysis of Roman graffiti follows Benefiel, "Dialogues of Ancient Graffiti." The circulation of letters within the early Christian church is described by Haines-Eitzen, *Guardians of Letters*; Thompson, "The Holy Internet"; and Klauck and Bailey, *Ancient Letters and the New Testament*.

## CHAPTER 3: HOW LUTHER WENT VIRAL

The advent of the printing press is described in Clair, *A History of European Printing*; Eisenstein, *The Printing Press As an Agent of Change* and *Divine Art, Infernal Machine*; Febvre and Martin, *The Coming of the Book*; and Pettegree, *The Book in the Renaissance*. The account of the viral spread of Luther's pamphlets and their subsequent impact draws chiefly upon Edwards, *Printing, Propaganda, and Martin Luther*; Scribner, *For the Sake of Simple Folk* and "Oral Culture and the Diffusion of Reformation Ideas"; Hillerbrand, *The Reformation in Its Own Words*; and Pettegree, *Reformation and the Culture of Persuasion*.

## CHAPTER 4: POETRY IN MOTION

The Devonshire Manuscript is available in full in a special "social edition" at http://en.wikibooks.org/wiki/The_Devonshire_Manuscript and its editors describe the social networks encapsulated in the book in "Drawing Networks in the Devonshire Manuscript." Its users, along with Thomas Wyatt, are the subject of Shulman, *Graven with Diamonds*. The circulation of manuscripts in the age of print and the practice of commonplacing are described by Love, *Scribal Publication*; Love and Marotti, "Manuscript Transmission and Circulation"; Carlson, "Manuscripts after Printing"; and Root, "Publication Before Printing." The account of the circulation of poetry for self-advancement follows Kilroy, *The Epigrams of Sir John Harington*; Doelman, "Circulation of the Late Elizabethan and Early Stuart Epigram"; Pebworth, "John Donne, Coterie Poetry, and the Text as Performance"; Wilcox, "Informal Publication of Late Sixteenth-Century Verse Satire"; and Wollman, "The 'Press and the Fire.'"

## CHAPTER 5: LET TRUTH AND FALSEHOOD GRAPPLE

For the tale of John Stubbs and his hand, see Woudhuysen, *Sir Philip Sidney and the Circulation of Manuscripts* and Stubbs, *Gaping Gulf*. The account of the history of efforts to censor the press draws upon McElligott, *Royalism, Print and Censorship in Revolutionary England*; Clegg, *Press Censorship in Jacobean England*; Dooley and Baron, *The Politics of Information in Early Modern Europe*; and Bently and Kretschmer, *Primary Sources on Copyright (1450–1900)*. The circulation of news in manuscript form is discussed in Love, *Scribal Publication*, and Raymond, *News Networks in Seventeenth Century Britain and Europe*. For the satirizing of corantos, their compilers and readers, see Chartier, *Inscription and Erasure*. The account of the explosion of 1641, and the many pamphlet forms that emerged, follows Raymond, *Pamphlets and Pamphleteering in Early Modern Britain*. The origins and impact of Milton's *Areopagitica* are discussed by Raymond, *Pamphlets and Pamphleteering in Early Modern Britain*; Morehouse, *Areopagitica*; and Shuger, *Censorship and Cultural Sensibility*.

## CHAPTER 6: AND SO TO THE COFFEEHOUSE

The rise of the coffeehouse is discussed in Pincus, "Coffee Politicians Does Create"; Cowan, "Mr. Spectator and the Coffeehouse Public Sphere"; and my previous book *A History of the World in 6 Glasses*. Efforts to suppress the coffeehouses are the subject of Cowan, "The Rise of the Coffeehouse Reconsidered." For the use of coffeehouses by scientists, see Stewart, "Other Centres of Calculation." For Oldenburg, correspondence societies, and the origins of the *Philosophical Transactions*, see Webster, "New Light on the Invisible College"; Johns, "Miscellaneous Methods"; Stimson, "Hartlib, Haak and Oldenburg"; Lux and Cook, "Closed Circles or Open Networks?"; Kronick, "The Commerce of Letters"; and Schneider, *The Culture of Epistolarity*.

## CHAPTER 7: THE LIBERTY OF PRINTING

For early American newspapers see Stephens, *A History of News*, and Starr, *The Creation of the Media*. The circulation of news in colonial America, by clergymen in particular, is discussed in Brown, "Spreading the Word"; McIntyre, "'I Heare It So Variously Reported'"; and Kielbowicz, "Newsgathering by Printers' Exchanges Before the Telegraph." The National Humanities Center's "Becoming American" website provides primary sources for the disputes over the Stamp Act and Townshend Acts. The account of the social circulation of revolutionary ideas via pamphlets, letters, and newspapers draws upon Ponder, *American Independence*, and Calkin, "Pamphlets and Public Opinion during the American Revolution." The story of *Common Sense* is told by Stephens, *A History of News*; Starr, *The Creation of the Media*; Burns, *Infamous Scribblers*; and Nelson, "Thomas Paine and the Making of *Common Sense*."

## CHAPTER 8: THE SENTINEL OF THE PEOPLE

The story of Théophraste Renaudot, the Craig Newmark of his day, is told by Richardson, "The Conferences of Théophraste Renaudot," and Beveridge, "A Seventeenth-Century Labour Exchange." The structure of the French media is discussed by Popkin, *Revolutionary News*. The account of the underground media and the Affair of the Fourteen follows Darnton, *Poetry and the Police* and *The Devil in the Holy Water*. The explosion of pamphlets and news-sheets is described by Popkin, "Pamphlet Journalism at the End of the Old Regime," and Gough, *The Newspaper Press in the French Revolution*. For the story of Condorcet and the

Cercle Social see Goodman, *The Republic of Letters*; Eisenstein, *Divine Art, Infernal Machine*; and Kates, *The Cercle Social.*

## CHAPTER 9: THE RISE OF MASS MEDIA

The introduction of steam printing by the London *Times* is discussed by Lee, *The Origins of the Popular Press in England*, and Lee, *The Daily Newspaper in America*. For American newspapers after independence and the launch of the New York *Sun*, see Stephens, *A History of News*; Starr, *The Creation of the Media*; and Huntzicker, *The Popular Press*. The account of the emergence of dedicated reporters follows Stephens, *A History of News*; Weaver, *The American Journalist*; and Schudson, *Discovering the News*. Newspaper exchanges are discussed by Kielbowicz, "News Gathering by Mail in the Age of the Telegraph." The rise of the telegraph is described in my previous book *The Victorian Internet*. For the advent of media moguls see Stephens, *A History of News*, and Starr, *The Creation of the Media*.

## CHAPTER 10: THE OPPOSITE OF SOCIAL MEDIA

The account of the early years of radio in the United States follows White, *United States Early Radio History*; Hanson, "The Original WWW"; and Wu, *The Master Switch*. The rise of national broadcasters in America, Britain, and Germany is described in White, *United States Early Radio History*; Crisell, *An Introductory History of British Broadcasting*; and Stephens, *A History of News*. The invention and introduction of television are described in MacDonald, *One Nation Under Television*, and Parks and Kumar, *Planet TV.*

## CHAPTER 11: THE REBIRTH OF SOCIAL MEDIA

The birth of ARPANET is described by Markoff, "An Internet Pioneer Ponders the Next Revolution," and Raz, "'Lo' and Behold." The account of the origin of TCP/IP draws on an interview with Vint Cerf, and the account of the origin of the web draws on interviews with Sir Tim Berners-Lee. The history of social networking is described by Boyd and Ellison, "Social Network Sites: Definition, History, and Scholarship." The account of the role of social media in Egypt's revolution draws upon an interview with Wael Ghonim. The best roundup of the intersection of politics and the internet is MacKinnon, *Consent of the Networked.*

## EPILOGUE: HISTORY RETWEETS ITSELF

For the cases for and against the power of Internet activism, see Shirky, *Here Comes Everybody*; Morozov, *The Net Delusion*; and Gladwell, "Small Change." A good synthesis is provided by MacKinnon, *Consent of the Networked.* Concerns about the antisocial effects of social media are expressed in Turkle, *Alone Together* and Powers, *Hamlet's BlackBerry*, and disputed in Tufecki, "Social Media's Small, Positive Role in Human Relationships." Evidence for the social benefits of social media is presented by Hampton, Sessions, and Her, "Core Networks, Social Isolation, and New Media," and Wellman, Garofalo, and Garofalo, "The Internet, Technology and Connectedness."

# Sources

Bainton, R. H. *Here I Stand: A Life of Martin Luther.* Hendrickson Publishers, 2009.

Baldwin, B. "The Acta Diurna." *Chiron* 9 (1979), 189–203.

Benefiel, R. R. "Dialogues of Ancient Graffiti in the House of Maius Castricius in Pompeii." *American Journal of Archaeology* 114, part 1 (2010): 59–102.

Bently, L. and M. Kretschmer, eds. *Primary Sources on Copyright (1450–1900).* Arts and Humanities Research Council: www.copyrighthistory.org.

Beveridge, W. H. "A Seventeenth-Century Labour Exchange." *Economic Journal* 24, no. 96 (December 1914): 635–636.

Blair, A. *Too Much to Know: Managing Scholarly Information Before the Modern Age.* New Haven: Yale University Press, 2010.

Boissier, G. *Tacitus: And Other Roman Studies.* New York: G. P. Putnam's Sons, 1906.

Bonwick, C. C. "An English Audience for American Revolutionary Pamphlets." *Historical Journal* 19, no. 2 (June 1976): 355–374.

Boyd, D. M. and N. B. Ellison. "Social Network Sites: Definition, History, and Scholarship." *Journal of Computer-Mediated Communication* 13, no. 1 (2007), http://jcmc.indiana.edu/vol13/issue1/boyd.ellison.html.

Bridges, R., ed. *The BBC's Recommendations for Pronouncing Doubtful Words: Broadcast English, 1929*, http://www.bl.uk/learning/timeline/item126866.html.

Briggs, A. and P. Burke. *A Social History of the Media from Gutenberg to the Internet.* Cambridge: Polity, 2010.

Brown, R. D. "Spreading the Word: Rural Clergymen and the Communication Network of 18th-Century New England." *Proceedings of the Massachusetts Historical Society* 94 (1982): 1–14.

Burns, E. *Infamous Scribblers: The Founding Fathers and the Rowdy Beginnings of American Journalism*. New York: Public Affairs, 2006.

Byrne, R. W. and N. Corp. "Neocortex Size Predicts Deception Rate in Primates." *Proceedings of the Royal Society of London B* 271 (2004): 1693–1699.

Calkin, H. L. "Pamphlets and Public Opinion during the American Revolution." *Pennsylvania Magazine of History and Biography* 64, no. 1 (January 1940): 22–42.

Carlson, D. "Manuscripts after Printing." In *Prestige, Authority and Power in Late Medieval Manuscripts and Text (York Manuscripts Conference)*. York: York Medieval Press, 2000.

Cartwright, K. *A Companion to Tudor Literature*. Oxford: Wiley-Blackwell, 2010.

Chartier, R. *Inscription and Erasure: Literature and Written Culture from the Eleventh to the Eighteenth Century*. Philadelphia: University of Pennsylvania Press, 2007.

Chisick, H. "Pamphlets and Journalism in the Early French Revolution: The Offices of the Ami du Roi of the Abbé Royou as a Center of Royalist Propaganda." *French Historical Studies* 15, no. 4 (Autumn 1988): 623–645.

Cicero, Marcus Tullius. *Letters to Atticus*. Edited and translated by D. R. Shackleton Bailey. Cambridge: Harvard University Press, 2001.

Clair, C. *A History of European Printing*. London: Academic Press, 1976.

Clark, W., J. Golinski, and S. Schaffer, eds. *The Sciences in Enlightened Europe*. Chicago: University of Chicago Press, 1999.

Clarke, B. *From Grub Street to Fleet Street: An Illustrated History of English Newspapers to 1899*. Aldershot: Ashgate, 2004.

Clegg, C. S. *Press Censorship in Jacobean England*. Cambridge: Cambridge University Press, 2001.

Cole, R. G. "Reformation Printers: Unsung Heroes." *Sixteenth Century Journal* 15, no. 3 (Autumn 1984): 327–339.

Condorcet, Jean-Antoine-Nicolas de Caritat. *Outlines of an Historical View of the Progress of the Human Mind*. London: J. Johnson, 1795.

Costolo, D. "The Power of Twitter as a Communication Tool," speech, Gerald R. Ford School of Public Policy, November 2012, http://www.fordschool.umich.edu/video/newest/1975704207001/.

Cowan, B. "Mr. Spectator and the Coffeehouse Public Sphere." *Eighteenth-Century Studies* 37, no. 3 (2004): 345–366.

———. "The Rise of the Coffeehouse Reconsidered." *Historical Journal* 47, no. 1 (March 2004): 21–46.

Crisell, A. *An Introductory History of British Broadcasting*. New York: Routledge, 1997.

Darnton, R. *The Devil in the Holy Water or the Art of Slander from Louis XIV to Napoleon*. Philadelphia: University of Pennsylvania Press, 2010.

———. *Poetry and the Police: Communication Networks in Eighteenth-Century Paris.* Cambridge: Belknap Press of Harvard University Press, 2010.

———. "Mademoiselle Bonafon and the Private Life of Louis XV: Communication Circuits in Eighteenth-Century France." *Representations* 87, no. 1 (Summer 2004): 102–124.

Dickens, A. G. *Martin Luther and the Reformation.* London: English Universities Press, 1967.

Doelman, J. "Circulation of the Late Elizabethan and Early Stuart Epigram." *Renaissance and Reformation* 29, no. 1 (2005): 59–73.

Dooley, B. M. and S. A. Baron. *The Politics of Information in Early Modern Europe.* New York: Routledge, 2001.

Druckenmiller, J. D. "Cicero's Letters and Roman Epistolary Etiquette." M.A. thesis, University of Oregon, December 2007.

Dunbar, R. I. M. "Coevolution of Neocortical Size, Group Size and Language in Humans." *Behavioral and Brain Sciences* 16, no. 4 (2003): 681–735.

———. "Gossip in Evolutionary Perspective." *Review of General Psychology* 8, no. 2 (2004): 100–110.

———. *Grooming, Gossip and the Evolution of Language.* Cambridge: Harvard University Press, 1997.

———. "Neocortex Size as a Constraint on Group Size in Primates." *Journal of Human Evolution* 20 (1992): 469–493.

———. "The Social Brain: Mind, Language, and Society in Evolutionary Perspective." *Annual Review of Anthropology* 32 (2003): 163–181.

———. "Theory of Mind and the Evolution of Language," in *Approaches to the Evolution of Language*, eds. Hurford, J. R., M. Studdert-Kennedy, and C. Knight. Cambridge University Press, 1998.

Edwards, C. "Epistolography," in *A Companion to Latin Literature*, ed. Harrison, S. Oxford: Blackwell, 2005.

Edwards, M. U. *Printing, Propaganda, and Martin Luther.* Berkeley: University of California Press, 1994.

Eisenstein, E. L. *Divine Art, Infernal Machine: The Reception of Printing in the West from First Impressions to the Sense of an Ending.* Philadelphia: University of Pennsylvania Press, 2011.

———. *The Printing Press as an Agent of Change: Communications and Cultural Transformations in Early-Modern Europe.* Cambridge: Cambridge University Press, 1979.

Engberg, E. "Blogging as Typing, not Journalism." CBSnews.com, November 8, 2004, http://www.cbsnews.com/stories/2004/11/08/opinion/main654285.shtml.

Epstein, Gady. "A Giant Cage: A Special Report on China and the Internet." *Economist*, April 6, 2013.

Farrell, J. M. "New England's Cicero: John Adams and the Rhetoric of Conspiracy." *Proceedings of the Massachusetts Historical Society* 104 (1992): 55–72.

Febvre, L. and H.-J. Martin. *The Coming of the Book: The Impact of Printing 1450–1800*. London: Verso Books, 1976.

Finlayson, J. G. *Habermas: A Very Short Introduction*. Oxford: Oxford University Press, 2005.

Fleming, Juliet. *Graffiti and the Writing Arts of Early Modern England*. London: Reaktion Books, 2001.

Foster, E. K. "Research on Gossip: Taxonomy, Methods, and Future Directions." *Review of General Psychology* 8, no. 2 (2004): 78–99.

Gibson, T. "Epilogue to Plato: The Bias of Literacy." *Proceedings of the Media Ecology Association* 6 (2005).

Giffard, C. A. "Ancient Rome's Daily Gazette." *Journalism History* 2 (1975): 106–132.

Gladwell, M. "Small Change: Why the Revolution Will Not Be Tweeted." *New Yorker*, October 4, 2010.

Goebbels, J. "Der Rundfunk als achte Großmacht." *Signale der neuen Zeit. 25 ausgewählte Reden von Dr. Joseph Goebbels*. Munich: Zentralverlag der NSDAP, 1938.

Goodman, D. *The Republic of Letters: A Cultural History of the French Enlightenment*. Ithaca: Cornell University Press, 1994.

Gough, H. *The Newspaper Press in the French Revolution*. London: Routledge, 1988.

Habermas, J. *The Structural Transformation of the Public Sphere: An Inquiry into a Category of Bourgeois Society*. Cambridge: MIT Press, 1989.

Haines-Eitzen, K. *Guardians of Letters: Literacy, Power and the Transmitters of Early Christian Literature*. Oxford: Oxford University Press, 2000.

Hampton, K. N., L. F. Sessions, and E. J. Her. "Core Networks, Social Isolation, and New Media: How Internet and Mobile Phone Use is Related to Network Size and Diversity." *Information, Communication & Society* 14, no. 1 (2011).

Hanson, W. "The Original WWW: Web Lessons from the Early Days of Radio." *Journal of Interactive Marketing* 12, no. 3 (1998): 46–56.

Harris, W. V. *Ancient Literacy*. Cambridge: Harvard University Press, 1989.

Herman, P. C. *Rethinking the Henrician Era: Essays on Early Tudor Texts and Contexts*. Chicago: University of Illinois Press, 1994.

Hillerbrand, H. J. *Christendom Divided: The Protestant Reformation*. London: Hutchinson, 1971.

———. *The Reformation in Its Own Words*. London: SCM Press, 1964.

Huntzicker, W. *The Popular Press, 1833–1865*. Westport: Greenwood Press, 1999.

Johns, A. "Miscellaneous Methods: Authors, Societies and Journals in Early Modern England." *British Journal for the History of Science* 33, no. 2 (June 2000): 159–186.

Jones, P. M. *Reform and Revolution in France: The Politics of Transition, 1774–1791*. Cambridge: Cambridge University Press, 1995.

Kaestle, C. F. "The History of Literacy and the History of Readers." *Review of Research in Education* 12 (1985): 11–53.

Kates, G. *The Cercle Social, the Girondins, and the French Revolution*. Princeton: Princeton University Press, 1985.

Kielbowicz, R. B. "News Gathering by Mail in the Age of the Telegraph: Adapting to a New Technology." *Technology and Culture* 28, no. 1 (January 1987): 26–41.

———. "Newsgathering by Printers' Exchanges before the Telegraph." *Journalism History* 9, no. 2 (Summer 1982): 42–48.

Kilroy, G. *The Epigrams of Sir John Harington*. Farnham: Ashgate Publishing, 2009.

Klauck, H.-J. and D. Bailey. *Ancient Letters and the New Testament: A Guide to Context and Exegesis*. Waco: Baylor University Press, 2006.

Kochan, D. J. "The Blogosphere and the New Pamphleteers." *Nexus Law Journal* 11 (2006): 99–109.

Kronick, D. A. "The Commerce of Letters: Networks and 'Invisible Colleges' in Seventeenth- and Eighteenth-Century Europe." *Library Quarterly* 71, no. 1 (January 2001): 28–43.

Lee, A. J. *The Origins of the Popular Press in England, 1855–1914*. London: Croom Helm, 1976.

Lee, A. M. *The Daily Newspaper in America: The Evolution of a Social Instrument*. New York: Macmillan, 1937.

Lewis, S. *News and Society in the Greek Polis*. Chapel Hill: University of North Carolina Press, 1996.

Love, H. *Scribal Publication in Seventeenth-Century England*. Oxford: Clarendon Press, 1993.

Love, H. and A. Marotti. "Manuscript Transmission and Circulation," in *The Cambridge History of Early Modern English Literature*, eds. Loewenstein, D. and J. Mueller. Cambridge: Cambridge University Press, 2002.

Lux, D. S. and H. J. Cook. "Closed Circles or Open Networks? Communicating at a Distance during the Scientific Revolution." *History of Science* 36 (1998): 179–211.

MacDonald, J. F. *One Nation Under Television: The Rise and Decline of Network TV*. New York: Pantheon Books, 1990.

MacKinnon, R. *Consent of the Networked: The Worldwide Struggle for Internet Freedom*. New York: Basic Books, 2012.

Markoff, J. "An Internet Pioneer Ponders the Next Revolution." *New York Times*, December 20, 1999, http://partners.nytimes.com/library/tech/99/12/biztech/articles/122099outlook-bobb.html.

Marotti, A. *Manuscript, Print, and the English Renaissance Lyric*. Ithaca: Cornell University Press, 1995.

McAndrew, Frank. "The Science of Gossip: Why We Can't Stop Ourselves." *Scientific American Mind*, October 2008.

McElligott, J. *Royalism, Print and Censorship in Revolutionary England.* Woodbridge, Suffolk: Boydell Press, 2007.

McIntyre, S. " 'I Heare It So Variously Reported': News-Letters, Newspapers, and the Ministerial Network in New England, 1670–1730." *New England Quarterly* 71, no. 4 (December 1998): 593–614.

Merrick, J. "Sexual Politics and Public Order in Late Eighteenth-Century France: The Mémoires Secrets and the Correspondance Secrète." *Journal of the History of Sexuality* 1, no. 1 (July 1990): 68–84.

Mierow, C. C. "Julius Caesar as a Man of Letters." *Classical Journal* 41, no. 8 (May 1946): 353–357.

Mithen, S. *The Prehistory of the Mind: A Search for the Origins of Art, Religion and Science.* London: Thames & Hudson, 1996.

Morehouse, I. M. *Areopagitica: Milton's Influence on Classical and Modern Political and Economic Thought.* Ludwig von Mises Institute, 2009, mises .org/daily/3738/.

Morozov, E. *The Net Delusion: The Dark Side of Internet Freedom.* New York: PublicAffairs, 2011.

Murphy, T. "Cicero's First Readers: Epistolary Evidence for the Dissemination of His Works." *Classical Quarterly* 48, no. 2 (1998): 492–505.

Myers, R. and M. Harris, eds. *Spreading the Word: The Distribution Networks of Print 1550–1850.* Winchester: St. Paul's Bibliographies, 1990.

National Humanities Center. "Becoming American: The British Atlantic Colonies, 1690–1793," http://nationalhumanitiescenter.org/pds /becomingamer/.

Nelson, C. "Thomas Paine and the Making of *Common Sense.*" *New England Review* 27, no. 3 (2006): 228–250.

Ong, W. J. *Orality and Literacy: The Technologizing of the Word.* London: Methuen, 1982.

Parks, L. and S. Kumar, eds. *Planet TV: A Global Television Reader.* New York: New York University Press, 2003.

Pauli, A. F. "Letters of Caesar and Cicero to Each Other." *Classical World* 51, no. 5 (February 1958): 128–132.

Peacey, J. *Politicians and Pamphleteers: Propaganda during the English Civil Wars and Interregnum.* Aldershot: Ashgate, 2004.

Pebworth, T.-L. "John Donne, Coterie Poetry, and the Text as Performance" *Studies in English Literature, 1500–1900* 29, no. 1 (Winter 1989): 61–75.

Pelli, D. G. and C. Bigelow. "A Writing Revolution." Seedmagazine .com, October 20, 2009, http://seedmagazine.com/content/article/a _writing_revolution/.

Pettegree, A. *The Book in the Renaissance.* New Haven: Yale University Press, 2010.

———. *Reformation and the Culture of Persuasion*. Cambridge: Cambridge University Press, 2005.

Pettegree, A. and M. Hall. "The Reformation and the Book: A Reconsideration." *Historical Journal* 47, no. 4 (December 2004): 785–808.

Pincus, S. "'Coffee Politicians Does Create': Coffeehouses and Restoration Political Culture." *Journal of Modern History* 67, no. 4 (December 1995): 807–834.

Ponder, B. *American Independence: From* Common Sense *to the Declaration*. Charleston: Estate Four Publishers, 2010.

Popkin, J. D. "Pamphlet Journalism at the End of the Old Regime." *Eighteenth-Century Studies* 22, no. 3 (Spring 1989): 351–367.

———. "The Press and the French Revolution after Two Hundred Years." *French Historical Studies* 16, no. 3 (Spring 1990): 664–683.

———. *Revolutionary News: The Press in France, 1789–1799*. Durham: Duke University Press, 1990.

Potter, S. "Webs, Networks and Systems: Globalization and the Mass Media in the Nineteenth- and Twentieth-Century British Empire." *Journal of British Studies* 46 (2007): 630.

Powers, W. *Hamlet's BlackBerry: A Practical Philosophy for Building a Good Life in the Digital Age*. New York: Harper Perennial, 2010.

Raymond, J. *News Networks in Seventeenth Century Britain and Europe*. London: Routledge, 2006.

———. *Pamphlets and Pamphleteering in Early Modern Britain*. Cambridge: Cambridge University Press, 2003.

Raz, Guy. "'Lo' and Behold: A Communication Revolution." *All Things Considered*, NPR, October 29, 2009, http://www.npr.org/templates/story/story.php?storyId=114280698.

Richardson, L. M. "The Conferences of Théophraste Renaudot: An Episode in the Quarrel of the Ancients and Moderns." *Modern Language Notes* 48, no. 5 (May 1933): 312–316.

Root, R. K. "Publication before Printing." *PMLA* 28, no. 3 (1913): 417–431.

Schmandt-Besserat, D. *How Writing Came About*. Austin: University of Texas Press, 1996.

Schneider, G. *The Culture of Epistolarity: Vernacular Letters and Letter Writing in Early Modern England, 1500–1700*. Newark: University of Delaware Press, 2005.

Schudson, M. *Discovering the News: A Social History of American Newspapers*. New York: Basic Books, 1978.

Scribner, R. W. "Oral Culture and the Diffusion of Reformation Ideas," in *Popular Culture and Popular Movements in Reformation Germany*. London: Hambledon Press, 1987.

———. *For the Sake of Simple Folk: Popular Propaganda for the German Reformation*. Cambridge: Cambridge University Press, 1981.

Shirky, C. *Here Comes Everybody: The Power of Organizing without Organizations.* New York: Penguin Press, 2008.

Shuger, D. K. *Censorship and Cultural Sensibility: The Regulation of Language in Tudor-Stuart England.* Philadelphia: University of Pennsylvania Press, 2006.

Shulman, N. *Graven with Diamonds: The Many Lives of Thomas Wyatt.* London: Short Books, 2011.

Siemens, R., J. Paquette, K. Armstrong, C. Leitch, B. D. Hirsch, E. Haswell, and G. Newton. "Drawing Networks in the Devonshire Manuscript (BL Add 17492): Toward Visualizing a Writing Community's Shared Apprenticeship, Social Valuation, and Self-Validation." *Digital Studies/Le champ numérique* 1, No 1 (2009).

Stacy, C. C. "Getting Started Computing at the AI Lab." Working Paper 235, Massachusetts Institute of Technology Artificial Intelligence Laboratory, September 7, 1982.

Standage, T. *A History of the World in 6 Glasses.* New York: Walker & Company, 2005.

———. *The Victorian Internet: The Remarkable Story of the Telegraph and the Nineteenth Century's On-line Pioneers.* New York: Walker & Company, 1998.

Starr, P. *The Creation of the Media: Political Origins of Modern Communications.* New York: Basic Books, 2004.

Stephens, M. *A History of News.* New York: Oxford University Press, 2007.

Stewart, L. "Other Centres of Calculation, or, Where the Royal Society Didn't Count: Commerce, Coffee-Houses and Natural Philosophy in Early Modern London." *British Journal for the History of Science* 32, no. 2 (June 1999): 133–153.

Stimson, D. "Hartlib, Haak and Oldenburg: Intelligencers." *Isis* 31, no. 2 (April 1940): 309–326.

Stowers, S. K. *Letter Writing in Greco-Roman Antiquity.* Philadelphia: Westminster Press, 1989.

Stubbs, John. *Gaping Gulf: With Letters and Other Relevant Documents.* Edited by Lloyd E. Berry. Charlottesville: University Press of Virginia, 1968.

Suetonius. *The Twelve Caesars.* London: Penguin, 1989.

The Devonshire MS Editorial Group. *A Social Edition of the Devonshire MS (BL Add 17492),* http://en.wikibooks.org/wiki/The_Devonshire_Manuscript.

Thompson, M. B. "The Holy Internet: Communication between Churches in the First Christian Generation," in *The Gospels for All Christians: Rethinking the Gospel Audiences*, ed. Bauckham, R. Grand Rapids: Wm. B. Eerdmans Publishing Company, 1997.

Timperley, C. H. *A Dictionary of Printers and Printing.* London: H. Johnson, 1839.

Traister, D. "Reluctant Virgins: The Stigma of Print Revisited." *Colby Quarterly* 26, no. 2 (June 1990): 75–86.

Tufecki, Z. "Social Media's Small, Positive Role in Human Relationships." *Atlantic*, April 25, 2012. http://www.theatlantic.com/technology/archive/2012/04/social-medias-small-positive-role-in-human-relationships/256346/.

Turkle, S. *Alone Together: Why We Expect More from Technology and Less from Each Other.* New York: Basic Books, 2011.

Weaver, D. H. *The American Journalist: A Portrait of U.S. News People and Their Work.* Bloomington: Indiana University Press, 1986.

Webster, C. "New Light on the Invisible College: The Social Relations of English Science in the Mid-Seventeenth Century." *Transactions of the Royal Historical Society* 24 (1974): 19–42.

Wellman, B., A. Garofalo, and V. Garofalo. "The Internet, Technology and Connectedness." *Transition* 39, no. 4 (Winter 2009): 5–7, http://homes.chass.utoronto.ca/~wellman/publications/internet_tech_connectedness/Wellman-Garofalos.English.pdf.

White, T. H. *United States Early Radio History.* http://earlyradiohistory.us/.

Wicks, J. "Roman Reactions to Luther: The First Year (1518)." *Catholic Historical Review* 69, no. 4 (October 1983): 521–562.

Wilcox, J. "Informal Publication of Late Sixteenth-Century Verse Satire." *Huntington Library Quarterly* 13, no. 2 (February 1950): 191–200.

Winsbury, R. *The Roman Book.* London: Duckworth Publishing, 2009.

Wollman, R. "The 'Press and the Fire': Print and Manuscript Culture in Donne's Circle." *Studies in English Literature, 1500–1900* 33, no. 1 (Winter 1993): 85–97.

Woudhuysen, H. R. *Sir Philip Sidney and the Circulation of Manuscripts, 1558–1640.* Oxford: Clarendon Press, 1996.

Wright, R. *The Evolution of God.* London: Little, Brown, 2009.

Wu, T. *The Master Switch: The Rise and Fall of Information Empires.* New York: Alfred A. Knopf, 2010.

Zaret, D. *Origins of Democratic Culture: Printing, Petitions, and the Public Sphere in Early-Modern England.* Princeton: Princeton University Press, 2000.

# INDEX

abbreviations, 22, 182–83, 221
ABC (American Broadcasting Co.), 212
*Academia* (Cicero), 35
*acta diurna* (the "daily acts"), 2, 28–33
Adams, John, 135, 138–39, 145, 151, 168–69, 172
Adams, Samuel, 141
Addison, Joseph, 113
Advanced Research Projects Agency (ARPA), 214–16
advertisements
  on blogs, 228
  in "Feuilles du Bureau d'Adresse," 147–48
  in graffiti, 38, 39
  in newspapers, 173–76, 180
  on radio broadcasts, 199, 202, 205
  on social-networking sites, 248, 249
  and television, 209–10, 211, 212, 213
"Affair of the Fourteen, The," 154–56
Albrecht, Archbishop of Mainz, 51–52, 61
Aleander, 59
Al Jazeera, 235
"Amateur Radio Restored" (Gernsback), 195, 196
American Broadcasting Co. (ABC), 212
"American Crisis, The" (Paine), 145–46
*American Nervousness* (Beard), 184
American Revolution, 138–39, 140–46, 145. *See also* colonial North America
American Telegraph Co., 183
*Ami du peuple* (newspaper), 162
*Amusing Ourselves to Death* (Postman), 213
Andreessen, Marc, 223, 224
Antoinette, Marie, 167–68
AP (Associated Press), 184
Apollonius of Perga, 18
"Apology for Printers" (Franklin), 103
Arab coffee and coffeehouses, 105
Arab Spring, 234–37, 241–42
Archimedes, 18
*Areopagitica* (Milton), 98–103, 160, 244
Aristotle, 19
ARPA (Advanced Research Projects Agency), 214–16
ARPANET, 215–16, 217–21
Associated Press (AP), 184
Atkyns, Richard, 94
*Atlantic Monthly*, 184–85
AT&T, 199–200
Atticus, 26, 35, 36
Aubrey, John, 109

Baird, John Logie, 205, 206–7, 208
Barger, Jorn, 225
Bartlett, Josiah, 143
Bastwick, John, 86
BBC (British Broadcasting Corp.), 200–201, 207, 209, 212
Beard, George, 184
behavioral roots of social media systems, 7–8, 14
Belarus, 242–43
Ben Ali, Zine El Abidine, 235
Bennett, James Gordon, 175, 177, 179
Berkeley, William, 125

Berners-Lee, Tim, 123, 222–24, 249
Biard Television, 207
Bibulus, Marcus Calpurnius, 28
Biddulph, William, 105
Bigelow, Charles, 239
blogosphere, 226
blogs and blogging
  about, 225–26, 228–29
  coffeehouses compared to, 123
  and mainstream media, 227
  media environment of 1640s compared to, 103
  microblogging, 231–34, 237–39
  miscellanies compared to, 74
  services, 226
  topics, 226–27
Boleyn, Anne, 64–65, 70
Bonafon, Marie Madeleine Joseph, 158
Bonis, François, 154
*Book of the Courtier, The* (Castiglione), 79
books
  elaborations, 48–49
  miscellanies, 65–68, 73–77
  papyrus rolls, 34, 46
  *pecia* system, 49–50
  reviews and summaries, 119–20
  roman à clef, 158–59
  of Rome and Romans, 34–38
*Boston Daily Times* (newspaper), 175–76
*Boston Gazette* (newspaper), 128, 135, 144
*Boston News-Letter* (weekly), 127
Bouazizi, Mohamed, 234
Boylston, Zabdiel, 129
Bradford, William, 125
Branch, 234
Brissot, Jacques-Pierre, 103, 161, 162, 164, 165, 168
British Broadcasting Corp. (BBC), 200–201, 207, 209, 212
broadcast era
  audio transmissions, 196–99
  wireless, 189–96
  *See also* television
*Broadcasting* (Rothafel and Yates), 205–6
*Broadcast Over Britain* (Reith), 201
Brutus, Marcus, 35
Bureau d'Adresse et de Rencontre, Paris, 147–48
Burton, Henry, 86
Bush, George W., 227–28
Butler, Samuel, 108
Butter, Nathaniel, 90, 91
Caelius Rufus, Marcus, 32–33
Caesar, Julius, 1–2, 23, 27, 28–29
Cambridge University, 111–12
Campbell, John, 126–28
Carroll, Charles, 135
Castaing, Jonathan, 116
Castiglione, 79
Catholic Church, 51–53, 56, 59–62, 85, 100
CBS (Columbia Broadcasting System), 200, 210
censorship
  in China, 159, 237, 243
  in France, 149, 160
  Milton on, 98–101
Cercle Social (social circle), 165–66
Cerf, Vinton, 219
Chamberlain, John, 87–88, 90
"Character of a Coffee-house, The" (pamphlet), 112
Charles I, 91–95, 101
Charles II, 102, 108
chat rooms, telegraph compared to, 182, 183
Chaucer, 68–69
China, 159, 233, 237–38, 243
China's microblogs (*weibo*), 233, 237–38
Christianity, 42–49, 51–53, 56, 61, 62, 74. *See also* Catholic Church; Luther, Martin
Church, Benjamin, 134
Church of England, 85–86, 98
Cicero, Marcus Tullius
  and *acta diurna*, 31, 32–33
  book-writing, 35–36
  communication system, 1–3, 246
  correspondence of, 22–23, 24, 26–28, 31, 113
  political goal, 29
clay tablets, 15–16
Clement, 44
codex format, 46–47
coffee, 105
coffeehouses
  about, 104, 110, 113
  complaints about, 111–12, 245
  discussion subjects, 106–8, 108–9, 111, 114–17
  regulation attempt, 108
  and scientific journals, 123
  spread of, in Europe, 105–6
  Twitter compared to, 232–33
"Coffee-houses Vindicated" (pamphlet), 113
Cohen, Jared, 242
colonial North America

British regulations, 133–38
newsletters of clergy, 125–27
newspapers, 127–28, 129–32, 133, 136–37
pamphlets, 126–27, 128–29, 131–32, 134–36
printing regulations, 125
Second Continental Congress, 141, 145
and Thomas Paine, 139–46
*See also* American Revolution
Columbia Broadcasting System (CBS), 200, 210
commonplace books (miscellanies), 65–68, 73–77
*Common Sense* (Paine), 140
computer networks
ARPANET, 217–21
e-mail systems, 215, 219–20, 229–30
IMPs, 216–17
interactivity of browsers and servers, 248–49
Internet, 219
Internet service providers, 221–22, 225
need for, 214–16
Netnews / Usenet, 221
packet switching protocol, 216–17
World Wide Web, 222–28
*See also* social-networking sites
Condorcet, Nicolas de, 164, 165, 166–67, 168, 242
*Connecticut Courant* (newspaper), 144
Cooke, Elisha, Jr., 128–29
corantos, 90–91
Cornificius, Quintus, 36
Cosby, William, 131–32
Costolo, Dick, 232–33
Cotton, John, Jr., 126
*Courier and Enquirer* (newspaper), 177
*Courier de l'Europe* (newspaper), 151
"Course of the Exchange and Other Things" (Castaing), 116
Cranach, Lucas, 58–59
Cromwell, Oliver, 101–2
Cromwell, Thomas, 70
cuneiform writing, 15–17
Cushing, William, 145

d'Angerville, Barthélemy-François Moufle, 157
Darnton, Robert, 159
Davies, Sir John, 80, 81
Day, Benjamin, 173–75, 177
"Deceiver Unmasked, The" (clergymen), 145
*De Finibus* (Cicero), 35
*Democratic Press* (newspaper), 177
Desmoulins, Camille, 162, 163
Devonshire Manuscript, 65–68
D'Ewes, Sir Simonds, 89
"Dialogue Against . . . Martin Luther" (Mazzolini), 55–56
Dickinson, John, 136, 137–38
Dionysius, 19, 44
"Discoverie of a Gaping Gulf . . ." (Stubbs), 82–83
"Disputation on . . . Indulgences" (Luther), 52
"Diurnall Occurrences . . . in Parliament" newsbooks, 92–95
divorce pamphlet, 98
Donne, John, 77, 80–81
dotcom boom, 224–25
Doublet, Madame, 157
Douglas, Lady Margaret, 64–65, 67–69
Dudley, Joseph, 128
Dudley, Paul, 128–29
Dulany, Daniel, 135
Dunbar number, 11–14
Dunbar, Robin, 10–11, 14
Duvall, Bill, 217

Eck, Johann, 56
Edict of Worms, 60–61
Egypt, 16–17, 235–37, 242
*Electrical Experimenter* magazine, 192, 195
*Electrical World* magazine, 191–92
*Electrician and Mechanic* magazine, 191, 193–94, 196
Electro Importing Co. (Telimco), 189–90
Elizabethan era, 77–81, 82–84
Elizabeth II, 212
Ellis, Jim, 221
e-mail systems, 215, 219–20, 229–30
Engberg, Eric, 227
England
BBC, 200–201, 207, 209, 212
Church of England, 85–86, 98
media explosion in 1640s, 92–98
Milton on freedom of expression, 98–101, 102–3, 160
news in, 87–89, 92–98, 103
printer regulations and licensing, 83–86, 89–90, 97–98, 101–2
Royal Society, 114, 118, 119–23
stamp duty, 171–72
Thirty Years' War, 90, 91
*See also* coffeehouses; *specific rulers*

ENQUIRE software, 222
enterprise social networks, 245–46
Erasmus of Rotterdam, 62, 73–74, 84, 243–44
Europe and coffeehouses, 105
Evans, Harold, 228
Evelyn, John, 120–21

Facebook, 6–7, 12–13, 74, 230–34, 236
Facebook-like social network in the workplace, 245–46
Fama (goddess of rumor), 33
Farnsworth, Philo T., 207–8
Federal Communications Commission (FCC), 210
Federal Radio Commission (FRC), 200
Fertile Crescent, 15–17
"Feuilles du Bureau d'Adresse" (newsletter), 147–48
Flickr, 234
*florilegia* of Christians, 74
Fly, James, 210
forgery, 37–38
Forum in Rome, 2, 22, 28–29, 30–31
foundations of social sharing, 7–8
France
  Bureau d'Adresse et de Rencontre, 147–48
  Cercle Social, 165–66
  coffeehouses in, 106
  and free press, 163–69
  gossip in, 156–58, 159
  newspapers in, 149–51
  tax increases in 1780s, 160, 161
  underground social-media environment, 151–59
  *See also specific rulers*
Franklin, Benjamin, 103, 130–31, 132–33
Franklin, James, 130–31
FRC (Federal Radio Commission), 200
Frederick III of Saxony, 61
freedom of expression
  in colonial North America, 132
  drawbacks to, 167–69
  in France, 163–69
  impact of Internet, 242, 243
  Milton on, 98–101, 102–3, 160
  Mirabeau on, 160
freedom of the press, 172–73
French and Indian War, 133
French Revolution, 161–63, 164–65, 167–69, 242
friends groups at social-networking sites, 12–13, 229–30, 231, 246
Friendster, 229–30
Fröben, Johann, 57
Fuller, Thomas, 97
Fust, Johann, 51

Gaius, 45–46
Galen, 37–38
Galloway, John, 136
*Gazette de France* (newspaper), 148–50
*Gazette de Leyde* (newspaper), 150, 151, 161
*Gentlemen's Magazine*, 116
George III, 140
Germany, 202–5
Gernsback, Hugo, 189–90, 192, 195
Ghonim, Wael, 236, 242
Gladwell, Malcolm, 241
Goebbels, Joseph, 202, 204
Google and Google+, 234
gossip, 7–8, 13–14, 33, 87–89, 151–58
Graaf, Reinier de, 122
graffiti, 38–42
"Grand Concern of England Explained, The" (pamphlet), 112
Greece, literacy in, 17–20
Greek words in Roman letters, 22–23
Greeley, Horace, 186
grooming coalitions, 9–10, 13, 21–22
group living, 9–14
Gutenberg, Johann, 50–51

Habermas, Jürgen, 188
Hallaire, Jacques Marie, 156
Halley, Edmond, 114–15
Hamilton, Andrew, 132
*Hamlet's BlackBerry* (Powers), 246
Harington, Sir John, 78–80, 81
Harris, Benjamin, 124–25
hawkers (newsboys), 174
Hearst, William Randolph, 186–87
Helvius, Gnaeus, 39
Henry VIII, 57, 64–65, 70–71
Herculaneum, 38–39
Herschel, John, 174
hieroglyphic writing, 15–17
Hipparchus, 18
"His Majesties Answer to the Petition" (Charles I), 94
historical roots of social media systems, 7–8, 14
*History of the American Revolution, The* (Ramsay), 145
Hitler, Adolf, 202, 204–5
Hooke, Robert, 107, 113, 114–16

Hoover, Herbert, 199
House of Maius Castricius, 41
Howard, Henry, Earl of Surrey, 72
Howard, Lord Thomas, 64–65, 67–69
Hus, Jan, 61, 242
Huygens, Constantijn, 122
hypertext documents (web pages), 222

Ignatius of Antioch, 44–45
IMPs (Interface Message Processors), 216–17
ink of Romans, 24
"Insania" (Dudley), 128–29
Instagram, 234
intelligencers, 117, 118
Interface Message Processors (IMPs), 216–17
Internet, 219. *See also* computer networks
Internet service providers, 221–22, 225

James I, 88
James II, 125
Jefferson, Thomas, 172
Jenkins, Charles Francis, 205, 206, 208
Jewett, Helen, 179
Jiang, Min, 159
Johnson, Nicholas, 212–13
Jonson, Ben, 91
*Journal des Sçavans*, 118, 121
*Journal of Commerce* (newspaper), 177
Juvenal, 31–32

Karlstadt, Andreas, 56, 62
Kleinrock, Leonard, 214, 216–17
Kline, Charley, 214, 216–17
Koenig, Frederick, 171

language
  driving force for development of, 14
  gossip, 7–8, 13–14, 33, 87–89, 151–58
  of Thomas Paine, 140–41
  *See also* literacy; writing
Laud, William, archbishop of Canterbury, 85–86, 92
Lavicomterie, Louis-Charles de, 165
Lee, Charles, 143
Leeuwenhoek, Antoni van, 122
Leo X, Pope, 56
L'Estrange, Roger, 102
letter pamphlets, 96
letters
  about, 16
  of Cicero, 22–23, 24, 26–28, 31, 113
  in colonial North America, 125–27
  New Testament, 42–47
  of Romans, 22–24, 26–28
  of scientific community, 117–18
  *See also* mail delivery; newsletters
"Letters from a Farmer in Pennsylvania . . ." (Dickinson), 137–38
"Letter to the Saxon Princes" (Luther), 62
"Liberty and Property Vindicated . . ." (Church), 134
"Liberty of Conscience" (Robinson), 101
*librarius* (specialist scribe), 34–35
literacy
  in Ancient Greece, 17–20
  in Dark Ages, 48–49
  Egyptain scribal-training texts, 16–17
  and graffiti in Roman world, 41–42
  and miscellanies, 73
  news ballads for illiterate, 57–58
  and newspaper sales, 175
  and pamphlets, 54
  pride in, 25
  and vowels, 18
living in a group, 9–14
Lloyd's of London, 116
*London Gazette*, 126
London Stock Exchange, 116
London *Times*, 170–71, 206
Lott, Trent, 227
Louis XV, 158
Louis XVI, 167–68
Luther, Martin
  broadsheets/woodcuts, 58–59
  and Catholic church, 51–53, 56, 59–62
  impact of, 242
  news ballads, 57–58
  pamphlets, 52–57, 61–63
"Luther's Game of Heresy" (woodcut), 59

mail delivery
  at coffeehouses, 109
  Franklin's role as a postmaster, 132–33
  newspaper delivery subsidies, 172–73
  in Roman world, 2, 24–28, 43–44
Mainz, Germany, 51
Mairobert, Mathieu-François Pidansat de, 157
Maissemy, Poitevin de, 160
Manuel, Pierre, 151–52
Marat, Jean-Paul, 162, 167
Marconi, Guglielmo, 190–91, 192
Marconi Wireless Telegraph Co., 191, 197
"Martyrdom of Polycarp, The" (Gaius), 45–46
*Maryland Gazette* (newspaper), 136

mass media
about, 3–4, 188, 239
mass-produced news, 170–76
penny newspapers as, 173–76, 177, 179–80, 185–87
*See also* broadcast era; television
Mather, Cotton, 126, 128, 129
Maurepas, Comte de (Jean-Frédéric Phélypeaux), 149–50, 152–54, 159
Mazzolini, Sylvester, 55–56
McKinsey and Company, 245–46
Mead, Joseph, 89, 90
media moguls, 186–87
media technology
about, 3–5, 8
broadsheets/woodcuts, 58–59
clay tablets, 15–16
codex format, 46–47
corantos, 90–91
for gossip at a distance, 14
miscellanies, 65–68, 73–77
news ballads, 57–58
newsbooks, 92–95, 101–2, 120
newsletters, 125–27, 147–48, 156–58
papyrus, 1–3, 16, 23–24, 34, 46, 47
parchment, 46, 48–49, 50
*pecia* system, 49–50
scientific journals, 117, 119–23
telegraph, 180–85
transitional phases, 247, 249
wax tablets, 23, 24, 46
white noticeboard or *album*, 29–30
*See also* books; computer networks; letters; mass media; newsletters; newspapers; pamphlets
Medium, 234
"Memorial of the Present Deplorable State of New England, A" (Mather), 128
*Mercurius Aulicus* (Royalist newsbook), 94–95, 99
*Mercurius Britannicus* (parliamentarian newsbook), 95, 96
Mesopotamia, 15–17
"Metamorphosis of Ajax, The" (Harington), 80
microblogging, 231–34, 237–39
Milton, John, 98–103, 160, 244
Mirabeau, Comte de, 160, 161, 168
miscellanies, 65–68, 73–77
*Modern Electrics* magazine, 191–92
"Modest Enquiry into the Grounds and Occasions of a Late Pamphlet, A" (Dudley), 128
Moore, Maurice, 134–35
Morozov, Evgeny, 241, 242–43
Morse code, 181, 190–91, 196, 197
Morse, Samuel, 180–82
Mosaic browser, 223
Mubarak, Hosni, 236
Müntzer, Thomas, 62
Murdoch, Rupert, 230
Murner, Thomas, 60
Myconius, Friedrich, 53
MySpace, 7, 230–31

Nag Hammadi texts, 46
National Assembly, France, 161, 162, 163, 164–65
National Broadcasting Co. (NBC), 200
*Naturalis Historia* (Pliny the Elder), 30–31
Nazi Germany, 202–5
NBC (National Broadcasting Co.), 200
neocortex, 8–9, 10–11, 12
Neolithic period, 15–16
Netnews/Usenet, 221
Netscape Communications, 224
networks. *See* computer networks
*New-England Courant* (newspaper), 129–31
*New London Gazette* (newspaper), 144
Newman, Noah, 125–26
*New Orleans Gazette* (newspaper), 176
news ballads, 57–58
newsbooks, 92–95, 101–2, 120
"News from the Coffee-House" (poem), 105
newsgroups, Netnews, 221
newsletters
*Boston News-Letter*, 127
of clergy in colonial North America, 125–27
of coffeehouses, 116
in England, 87–91
"Feuilles du Bureau d'Adresse," 147–48
for gossip in France, 156–58
*See also* letters
newspaper advertisements, 173–76, 180
newspaper reporters, 176–80, 184–85
newspapers
about, 171, 174, 187–88
in colonial North America, 127–28, 129–32, 133, 136–37
delivery subsidies, 172–73
in France, 148–50, 151, 161–62
penny papers, 173–76, 177, 179–80, 185–87
*See also specific newspapers*

New Testament, 42–47, 61
Newton, Isaac, 114, 115
*New York Herald* (newspaper), 175, 179, 194
New York *Morning Journal* (newspaper), 186–87
New York *Sun* (newspaper), 173–75, 177
*New York Times* (newspaper), 175, 186, 237
*New York Transcript* (newspaper), 179–80
*New York Tribune* (newspaper), 180
*New-York Weekly Journal* (newspaper), 131–32
"Ninety-Five Theses" (Luther), 52–53, 62
non-social media, 3–4. *See also* mass media
*nouvellistes*, 156–58
"Now We Drive Out the Pope" (news ballad), 58

Oldenburg, Henry, 117–23
Onesimus and Secundus, 41
"On His Own Books" (Galen), 38
oratory, writing as threat to, 18–20
"Origin of the Monks, The" (woodcut), 58
"Origins of the Pope, The" (woodcut), 58
"Orlando Furioso" (Italian epic poem), 79–80
ostracism and literacy, 18
*Outlines of . . . the Human Mind* (Condorcet), 166–67, 168–69
Oxford University, 111

packet switching protocol, 216–17
Paine, Thomas, 139–46, 165–66
pamphlets
  and coffeehouses, 109, 112–13
  in colonial North America, 126–27, 128–29, 131–32, 134–36
  in France, 160–61, 162
  Luther's use of, 52–57, 61–63
  media explosion in 1640s England, 92–95, 97–98
  Milton's, 98–103, 160, 244
  Paine's, 140–46
  responding to, 95–97
papal infallibility doctrine, 59–60
paper, 49, 50, 73, 133–34, 175
papyrus, 1–3, 16, 23–24, 34, 46, 47
parchment, 46, 48–49, 50
Parker, Henry, 94
"Passional Christi und Antichristi" (woodcuts), 59
Path, 234
*Patriote français* (newspaper), 161, 162
Paul of Tarsus, 42–44, 56
*pecia* system, 49–50
peer review, 119
Pelli, Denis, 239
*Pennsylvania Evening Post* (newspaper), 144
*Pennsylvania Gazette* (newspaper), 131, 133, 136–37, 145
*Pennsylvania Journal*, 139
*Pennsylvania Magazine*, 139
*People's Daily Online* "Online Public Sentiment" report, 237
Pepys, Samuel, 111
*Père Duchesne*, 163
periodicals, 109
Perry, K. W., 193
person-to-person information sharing. *See* social media systems; social-networking sites
Peter, Apostle, 44
Petronius, 30
Pettegree, Andrew, 61
*Phaedrus* (Plato), 19, 246
*Philosophical Transactions of the Royal Society of London*, 120–23
Pinterest, 74, 234
Pionius, 46
"Plain Truth" (Candidus), 144–45
"Plain Truth" (Paine), 140
Plato, 19–20, 246
play pamphlets, 96–97
Pliny the Elder, 30–31
Pliny the Younger, 32, 37
poetry
  and coffeehouses, 106, 113
  Devonshire Manuscript, 65–68
  in Elizabethan era, 77–81
  French news as, 152–58
  manuscript networks, 72–77
  in miscellanies, 65–68, 73–77
  "News from the Coffee-House," 105
  in Tudor court, 69–72
*poissonnades*, 153–54
Polycarp, the bishop of Smyrna, 45–46
Pompadour, Madame de, 153–54
Pompeii, 25, 38–42
*Popular Science Monthly*, 199
Pory, John, 88, 89, 91
Postman, Neil, 213
Post Office Act (1792), 172–73
Powers, William, 246, 247
Preece, William, 190–91
primate brain/social brain evolution, 7, 8–11
*Principia* (Newton), 115
printers, 54, 57, 60, 163, 186–87

printing
  Cicero's speculation on, 31
  complaints of intellectuals, 243–44
  and French Revolution, 164–65
  licensing texts in colonial North America, 125
  licensing texts in England, 85–86, 89–90, 97–98, 101–2
  and Luther's campaign, 62–63
  Luther's view of, 61
  manuscript networks, 72–77
  printing press invention and improvements, 50–51, 170–71
  woodcuts, 50
  yearly increase in authors, 239
privilège system for French publishers, 148
"Pro Ligario" (Cicero), 36
Prynne, William, 86
*Publick Occurrences Both Forreign and Domestick* (newspaper), 124–25
Pulitzer, Joseph, 186

Quintilian, 37
"Qu'une bâtarde de catin" (song), 152

Radio Act (1912), 195
Radio Corporation of America (RCA), 197, 198–200, 207, 208–10, 212
radios and radio receivers, 197–99, 202
radio stations and tuning issues, 192–95
radio telegraphy, 190–95
radio telephony, 195–96
radio vision, 205–6
Ramsay, David, 145
rating businesses in Roman world, 40
RBC (Reich Broadcasting Corp.), 202–5
RCA (Radio Corporation of America), 197, 198–200, 207, 208–10, 212
reading. *See* literacy; scribes; writing
reciprocal personal relationships and group size, 12
*recitatio* (book launch party), 36–37
Reformation era
  broadsheets/woodcuts, 58–59
  dilemma for church, 59–60
  news ballads, 57–58
  pamphlets, 52–57, 61–63
  *See also* Luther, Martin
regulations
  of coffeehouses, 108
  controls on printers, 83–85
  in Elizabethan era, 82–84
  e-mail systems, 220
  licensing texts in colonial North America, 125
  licensing texts in England, 85–86, 89–90, 97–98, 101–2
  Milton on freedom of expression, 98–101, 102–3, 160
  Parliament's end to, 92–97, 244
  political news circumventing, 86–89
  prepublication approval of newspapers, 131
  privilège system for French publishers, 148
  of radio telephony, 200
  Stamp Act and colonial North America, 133–37
  stamp duty in England, 171–72
  Tea Act, 138
  Townshend Acts, 137–38
  for wireless telegraphy, 192–95
  workplace regulations on social-networking sites, 245–46
  *See also* censorship
Reich Broadcasting Corp. (RBC), 202–5
Reith, John, 200–201
Renaudot, Théophraste, 147–49
Righi, Augusto, 190
Robespierre, Maximilien de, 166, 168
Robinson, Henry, 101, 244
roman à clef (novel with a key), 158–59
Roman letter carriers (*tabellarii*), 24
Rome and Romans
  *acta diurna*, 2, 28–33
  books, 34–38
  and Christian's use of written documents, 42–48
  the Forum, 2, 22, 28–29, 30–31
  graffiti, 38–42
  letters and letter carriers, 22–28
  political structure, 28–29
  social media system, 1–3
  territories, 21
  *See also* scribes
Root, Elihu, 18, 147
Rosée, Pasqua, 105–6
Royal Society, 114, 118, 119–23
Rushworth, John, 97

Saerchinger, César, 202
Said, Khaled, 235–36
*San Francisco Call*, 192
*San Francisco Chronicle* (newspaper), 208
San Francisco *Examiner* (newspaper), 186
Sarnoff, David, 199–200, 207, 208

*Satires* (Juvenal), 31–32
*Satyricon* (Petronius), 30
Saultchevreuil, Étienne Lehodey de, 164
Sawiris, Naguib, 236–37
Schöffer, Peter, 51
*Scientific American* magazine, 182
scientific journals, 117, 119–23
screw press, 50
scribes, 16–17, 22, 23, 31, 34–35, 43, 45–46
scriptoria (copying rooms in monasteries), 48–49
Scudamore, Viscount, 89
*Seattle Daily Times* (newspaper), 196
Second Continental Congress, Philadelphia, 141, 145
Secundus and Onesimus, 41
Sedition Act, 172
Seneca, 26, 30, 246
"Sermon to Parents, A" (Gernsback), 189–90
Seventh Letter (Plato), 19
Seven Years' War, 133
Sewell, Samuel, 127–28
Shakespeare, William, 77
Shelton, Mary, 65, 67–68
Shirky, Clay, 241
Sidney, Philip, 77
"Sincere Admonition . . ." (Luther), 62
SixDegrees.com, 229
*60 Minutes*, 227–28
slavery, 22, 31, 36, 139
smallpox epidemic, 129
Smith, Adam, 117
Smith, William "Cato," 145
social brain evolution, 8–11
social circle (Cercle Social), 165–66
social grooming, 9–10
social media systems
  about, 3
  assessing and maintaining social position, 7–8, 13–14, 76–77
  behavioral and historical roots, 7–8, 14
  Internet as reassertion of, 4
  Luther's pamphlets, 53–57, 61–63
  manuscript networks, 72–77
  preindustrial media and digital media compared, 240–42
  and revolutions, 241–42
  synchronization of opinion, 62
  time spent engaging in, 244–46
  *See also* blogs and blogging; coffeehouses; computer networks; social-networking sites
social-networking sites
  about, 6–7, 247–50
  friends groups, 12–13, 229–30, 231, 246
  *weibo* (China's microblogs), 233, 237–38
  workplace regulations, 245–46
  *See also* blogs and blogging; *specific sites*
Spanish-American War, 186–87
*Spectator* (London magazine), 113
Speer, Albert, 204–5
Stamp Act, 133–37
Stamp Act Congress, 136
stamp duty in England, 171–72
"Staple of News, The" (Jonson), 91
Star Chamber of English king, 86, 92
Stevens, Thomas, 183
Stillman, W. J., 184–85
*Structural Transformation of the Public Sphere, The* (Habermas), 188
Stubbs, John, 82–84
Suetonius, 28
Sullivan, Andrew, 227, 228
synchronization of opinion, 62

*tabellarii* (Roman letter carriers), 24
Tacitus, 31
*Tatler* (London magazine), 107
Taylor, Bob, 214–16
TCP/IP protocol, 217, 219
Tea Act, 138
technology. *See* media technology
telegraphic news agencies, 184
telegraph system, 180–85
*Telephony* magazine, 196
television
  advertisements on, 210, 212, 213
  BBC, 200–201, 207, 209
  coach potatoes, 213
  color broadcasts, 212
  development of, 205–8
  Johnson's critique of, 212–13
  promotional advertisements, 209–10, 211
  RCA, 197, 198–200, 207, 208–10, 212
  social media as alternative to, 247
television receivers, 206, 207–8, 209, 210
Telimco (Electro Importing Co.), 189–90
Ten Rhijne, Willem, 122
*Terry* (torpedo boat destroyer), 193–94
*Test Pattern for Living* (Johnson), 212–13
Tetzel, Johann, 51–52, 55, 56
Thirty Years' War, 90, 91
Thomason, George, 95
*Titanic*, sinking of, 194
Tomlinson, Ray, 219

Townshend Acts, 137–38
train crash near Wenzhou, China, 238
"Troilus and Criseyde" (Chaucer), 68–69
Truscott, Tom, 221
Tufekci, Zeynep, 246–47
Tumblr, 74, 234
Tunisia, 234–35
Turkle, Sherry, 246
Twitter, 74, 232–34, 239, 243

United States. *See* American Revolution; colonial North America; computer networks
universities, 49, 247

Vail, Alfred, 181
Varro, Marcus Terentius, 35
Vieuxmaison, Madame de, 158
Virgil, 33–34
vowels, 18

Wallach, Dan, 237
War of Devolution, 120–21
Warren, James, 141, 143
Washington, George, 143, 145–46
wax tablets, 23, 24, 46
*Wealth of Nations, The* (Smith), 117
web browsers, 222–25
web pages (hypertext documents), 222
*Weekly Herald* (newspaper), 176
*weibo* (China's microblogs), 233, 237–38
Wenzhou, China, train crash, 238
white noticeboard or *album*, 29–30
Wight, John, 177
Winsbury, Rex, 22, 36
*Wired* magazine, 223
Wireless Association of America, 192
wireless broadcasting, 189–96
wireless telephony, 195–96, 198–99
wireless television, 207
Wisner, George, 177
Woodfall, William "Memory," 176
*World* (newspaper), 186
World War II, 209
World Wide Web, 222–28
"World Wide Wireless" (RCA logo), 197
Worshipful Company of Stationers, England, 97, 244
writing
- conversational, vernacular style, 95
- development of, 15–16, 36, 47
- in German vs. Latin, 52–53, 55, 61
- graffiti, 38–42
- hieroglyphic and cuneiform systems, 15–17
- scribes, 16–17, 22, 23, 31, 34–35, 43, 45–46
- and social brain evolution, 8
- telegraphic style, 184
- as threat to rhetoric, 18–20
- transitional form, 20
- *See also* abbreviations; letters; media technology; poetry

Wyatt, Sir Thomas, 65, 67–68, 70–72, 77
Wycliffe, John, 61, 242

"XII Resolves Concerning the Disposall of the Person of the King," 95–96

Young, Arthur, 160–61
YouTube, 234
YUMYUM restaurant guide, 219–20

Zenger, John Peter, 131–32
Zuckerberg, Mark, 230
Zworykin, Vladimir, 208

## A NOTE ON THE AUTHOR

Tom Standage is digital editor at *The Economist* and editor-in-chief of its website, Economist.com. He is the author of six history books, including *An Edible History of Humanity*, the *New York Times* bestseller *A History of the World in Six Glasses* and *The Victorian Internet*. His writing has also appeared in the *Daily Telegraph*, the *New York Times* and *Wired*. He lives in London.

tomstandage.com
@tomstandage
facebook.com/writingonthewallbook
instagram.com/tomstandage
flickr.com/photos/tomstandage

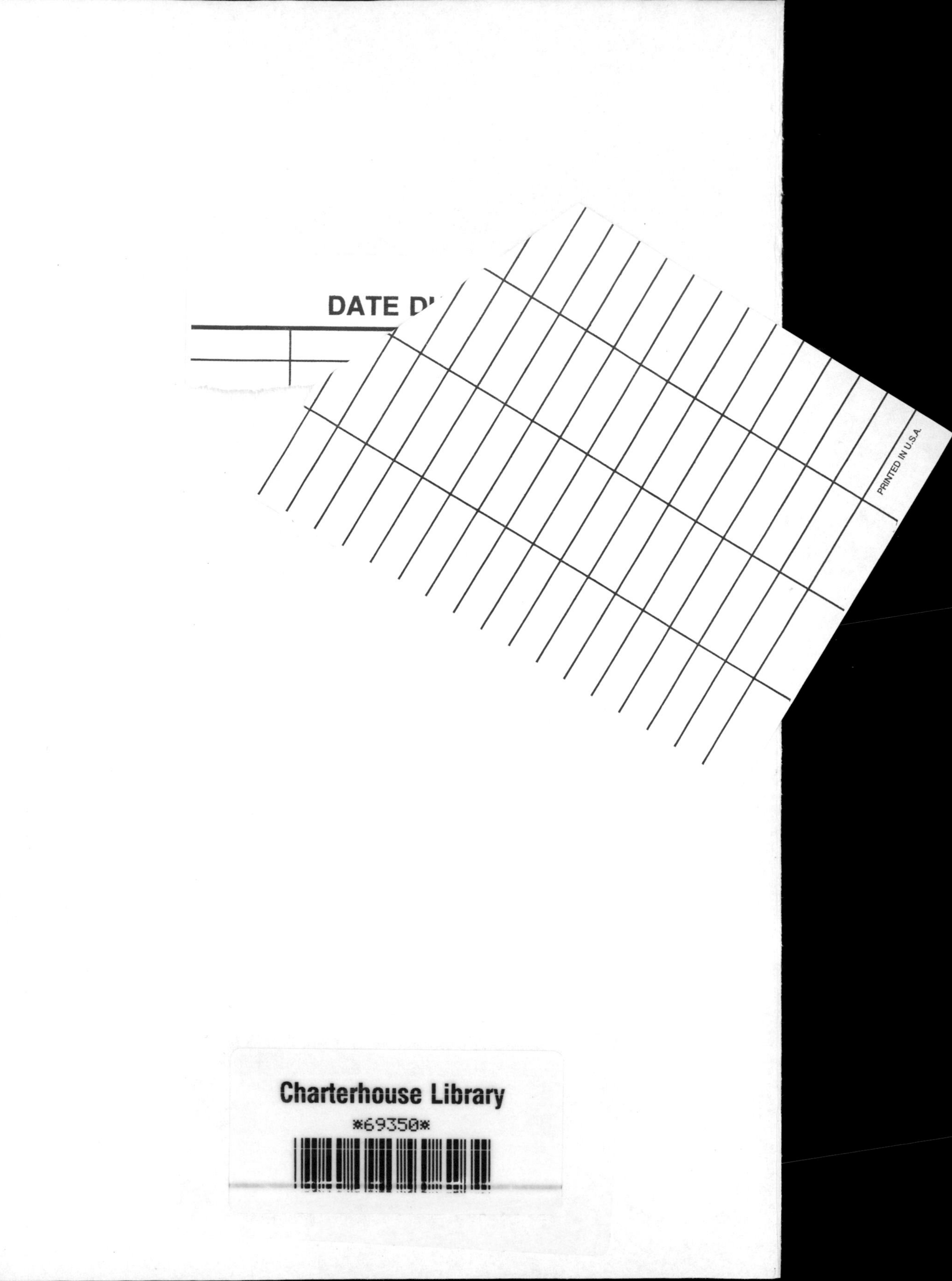
DATE D
PRINTED IN U.S.A.